RESTORING THE WILD

A GUIDE TO THE RESTORATION, CREATION AND MANAGEMENT OF MEADOW AND OTHER WILD VEGETATION

DONALD MacINTYRE

RESTORING THE WILD

A GUIDE TO THE RESTORATION, CREATION AND MANAGEMENT OF MEADOW AND OTHER WILD VEGETATION

Foreword by Sara Oldfield OBE

THE CROWOOD PRESS

CONTENTS

Foreword 7
Introduction 8

1 **Restoring the Wild** **11**
Wild Plants of Britain and Ireland 11
Restoration and Creation 14
The Vegetation Types Covered in this Book 16

2 **Biodiversity, Species, Ecotypes and Ecosystems** **19**
Biodiversity 19
Species, Ecotypes and Communities 20
Ecosystems 23

3 **Protected, Sensitive and Proscribed Species** **25**
Protected Species 25
Sensitive Species 27
Proscribed Species 28

4 **Invasive Alien Weeds and Native Weeds** **31**
Invasive Alien Weed Species 31
Native Weed Species 34

5 **Natural Capital and Ecosystem Services** **39**
Ecosystem Services 41

6 **Opportunities for Ecosystem Restoration and Creation** **47**
The Wild 48
Farmland 51
Gardens and Parks 56
City, Road and Rail 62
Wasteland and Derelict Land 62

7 **Seed Origin and Provenance** **65**
Native or Alien Species 66
Wild Type or Cultivar 66
British or Alien Ecotype 67
Local or Non-local Ecotype 67
Genetic Benefits and Threats 69

8 **Surveying, Planning and Methods** **73**
Botanical and Feature Surveys 73
Soil Type, Nutrient Status, Topography, Weeds and Stress 74
Opportunities for Natural Regeneration 76
Defining Project Objectives 77
Seed-based Methods 77
Plant-based Methods 77
Restoration, Natural Regeneration and Wilding 77
Programme of Work 79

9 **General Method for Meadow Restoration and Creation** **81**
General Method for Meadow Restoration 81
General Method for Meadow Creation 82
General Method for Other Habitats 82
Yellow Rattle (*Rhinanthus minor*) 83
Annual Nurse 86

10 Restoration Species and Mixture Composition **89**
Restoration Species 89
Mixture Composition 90
Annual Wild Flowers 90
Neutral Lowland Meadow and Pasture 91
Calcareous Meadow and Pasture 91
Vegetation on Sandy Soils 91
Floodplain, Wetland and Water Edge 92
Upland Meadow and Pasture 92
Moor and Heath 92
Wayside and Tussock 93
Hedgerow and Scrub 93
Woodland Ground Flora 93
Coastal 94
Orchids 95
General Purpose Mixture 95

11 Sourcing Seeds **97**
Hand-collected Wild Seed 97
Machine-harvested Meadow Seed 98
Crop-grown Native Wild Seed 103

12 Seed Processing and Seed Care **109**
Seed Processing 109
Seed Longevity and Storage 110
Seed Dormancy and Viability 111
Seed Germination 111

13 Ground Preparation and Seed Sowing **113**
Ground Preparation for Restoration Methods 113
Ground Preparation for Creation Methods 114
Seed Rates 114
When to Sow 114
How to Sow and Rolling in 114
First Signs of Germination 116

14 Management Methods and Tools **119**
Mowing and Cutting 119
Grazing 121
Weed Control 124

15 Habitat-specific Management with Examples **129**
Annuals 129
Meadow 130
Pasture 135
Wayside and Tussock 136
Floodplain, Wetland and Water Edge 138
Hedgerow and Scrub 140
Woodland 142
Orchids 144

16 Restoration Species Accounts **149**

Appendix 254
Glossary 261
References 264
Index 270

FOREWORD

The need for biodiversity conservation is ever increasing but policy and action are not keeping pace. This is frustrating for all who care about the environment. And it is deeply disappointing for those of us who started our environmental careers in the UK in the 1970s – a time when environmentalism was relatively new, exciting and optimistic. Of all the people who started their work in conservation at that time, I consider that Donald has arguably achieved the most! This book is a product of Donald's ambition, dedication and achievements over the past fifty years. It reflects a deep and practical knowledge of the British flora, and of how to manage landscapes and vegetation to retain and restore floral diversity. It is unique in bringing together scientific and applied information as a basis for the collection, production and utilisation of native seed – the wild material for ecological restoration. As noted in this book, since 1980 the annual production of native seed in the British Isles has increased a thousand fold. This is in no small part due to Donald's pioneering work in supplying good-quality native seed harvested directly from land grazed by shire horses, field-grown seed crops, and sympathetic collection from natural sites. Machinery such as the brush harvester has been developed by Donald and is now used worldwide in ecological restoration projects.

The production of good-quality native seed is enabling restoration and habitat creation to take place throughout the British Isles, supporting essential ecosystem services. The demand for native seed is growing. Donald's generosity in sharing his knowledge through this inspiring and practical publication will enable wider production of good-quality seed production to support and enhance biodiversity, whether in so-called wild areas, meadows, farmland, urban parks and gardens and along transport routes. The scope for further habitat creation and restoration using the methods described in this book remains huge. The production of a comprehensive list of restoration species and explanation of how and where they should be used is of great significance. It is remarkable that Donald has found the time to write such an important and hopeful book whilst still working full-time as a major native seed farmer, adviser on land management, and whilst ensuring that his family firm will continue to be the UK's largest producer of native seed for years ahead. With the increasing impacts of climate change, the importance of enhancing ecological resilience is ever more important and this book sets out how it can be achieved.

Sara Oldfield OBE, January 2024

INTRODUCTION

The Wild and the Cultivated

The wild are those places where the hand of nature, not a human hand, has had dominant control over the composition and functioning of ecosystems, and where nature is the provider of ecosystem services and the wild aesthetic. The cultivated are those places where the human hand, not the hand of nature, has dominant control over the composition of the vegetation. Nowhere is completely wild, or completely cultivated, but the concept of dominant control allows us to distinguish between *wild* and *cultivated* places, and we will use this definition throughout the coming pages, with restoration of the *wild* being our focus.

On the farm, the meadow is the mowing ground, pasture is the grazing land and arable is the plough land. Before agricultural intensification in the twentieth century, all three farmland uses, over most of Britain and Ireland, supported highly biodiverse, species-rich, wild herbaceous vegetation. By the late twentieth century 97 per cent of the meadows and calcareous grasslands had been converted to intensive arable farming and/or productive Ryegrass (*Lolium perenne*) swards. Likewise, a similar percentage of species-diverse lowland pasture and arable was lost to chemical farming. The surviving 3 per cent inspires a generation to bring back nature.

Looking way back in time, it is thought that our planet came into being over 4.6 billion years ago, that life originated in the oceans, or in fresh water, at least 3.5 billion years ago, and that the first land plants colonised dry land around 470 million years ago, during the Ordovician Period. At that time the CO_2 concentration in the atmosphere was around 0.5 per cent. It is now around 0.03 per cent. Since the Ordovician there have been four major Ice Ages, with lesser Ice Ages in between. The periodicity of Ice Ages is probably due in greater part to the ebb and flow of the carbon cycle. Photosynthesis in green plants, and plant-mediated chemical mineralisation of CO_2 (in sea and fresh water) contributes to the store of carbon in hydrocarbon deposits (such as coal, oil, peat) and mineralised carbonate deposits (chalk, limestone). As atmospheric CO_2 concentration declines (due to plant photosynthesis and mineralisation), the global mean temperature declines, and so dawns another Ice Age.

During each Ice Age the relative inactivity of plants triggers the carbon cycle to make a net positive contribution of CO_2 to the atmosphere, increasing atmospheric CO_2 concentration, thus causing temperatures to rise again. Higher CO_2 concentration and higher temperature increases the rate of photosynthesis, removing CO_2 from the atmosphere – and so is created the ebb and flow of glaciation. As ice sheets periodically recede, wild plants awaken, migrate, and prosper again across the planet (Lenton et al., 2012). In this way, and for millennia, wild plants have moderated global temperatures, climate and the composition of the atmosphere by buffering the carbon cycle. During this period of repeated glaciation, and before the advent of the Early Modern Human (*Homo sapiens*), around 300,000 years ago, everywhere and everything on Earth was completely wild. Now, nowhere and nothing on Earth is completely wild.

With each cycle of glaciation, land plants have migrated into and out of the region we now call the British Isles. The last glacial maximum was only about 22,000 years ago. At that time Britain was clothed in glaciers and glacial ice fields north of a line from the River Severn to the River Humber, with tundra/steppe vegetation to the south. The last glacial period ended about 10,000 years ago. Gradually temperature and sea levels rose, glaciers retreated, and wild temperate woodlands and wild temperate grasslands moved back in, crossing, we assume, the broad land bridge between Western Europe and Britain, replacing tundra and steppe and colonising the ice fields. The land bridge, Doggerland, was breached by rising sea levels about 8,000 years ago, and exists today as a submerged land form known as Doggerbank. This event cut off the British Isles and its vegetation from the rest of the European Continent. The vegetation at that time, and until the near present, probably consisted of native post-glacial temperate wild wood and wild grassland, in a landscape structure that may have resembled wood pasture (Peterken, 2013).

RECORDED LOSSES OF NATIVE WILD VEGETATION IN THE BRITISH ISLES

- 99.7 per cent loss of fen – 3,400sq km (1,300sq miles) – since 1637 (Thomas *et al*, 1981)
- 97 per cent loss of species-rich meadows and calcareous grasslands 1930–1984 (Fuller, 1987; Stephen, 1996)
- 81 per cent loss of grazing marsh from historic times to the near present (RSPB *et al*, 1997)
- 80 per cent loss of lowland heath 1800–1980 (Farrell, 1993)
- 44 per cent loss of lowland raised bog from historic times to the near present (Hume, 2008)
- 27 per cent loss of upland heath 1947–1980 (Bardgett *et al*, 1995)
- 7 per cent loss of ancient woodland *c.*1930–1985 (Spencer and Kirby, 1992)
- Over 40 per cent of wild species in the UK are currently in decline (Hayhow *et al*, 2019)

The flora of Britain and Ireland is one of the best studied floras worldwide. *The Plant Atlas 2020* (Stroh et al., 2023) is a huge credit to the thousands of field botanists who contributed to it, and it records an impoverished and declining native flora in Britain and Ireland when compared to earlier surveys. Little is left of the original post-glacial wild wood and wild grassland. Only fragmented and isolated refugia survive. The twentieth century, remarkable for human population growth and technological advance, may also be remembered as the period when we exploited, harmed, modified and polluted the environment in a completely unsustainable way. This has been little short of disastrous for nature in the British Isles (Bowers, 1985). Humans have left their mark. Land degradation resulting from cultivation and overgrazing, in the present interglacial, has reduced global photosynthetic capacity, perhaps to a point at which nature will struggle to bring down or moderate atmospheric CO_2 concentrations, as it has done in past interglacial periods.

The recent decades of sustained harm to nature have generated a low point that motivates all those engaged in nature restoration, as well as providing a baseline for nature recovery (Ridding et al., 2013). The tools of that recovery are the protection and beneficial management of existing wild vegetation, the restoration of depleted and damaged wild plant communities, and the creation of high-quality wild vegetation where none remains. The UK policy in support of nature recovery is outlined in the Lawton Report 'Making Space for Nature' (Lawton et al., 2010), the 25 Year Plan 'A Green Future' (UK Government, 2018), The Environment Act 2021 (UK Government, 2021) and the Environmental Improvement Plan (UK Government, 2023).

Author's Note

The plant names used in this book are those published in *The Concise Flora of The British Isles,* by Clive Stace (2022).

CHAPTER 1

RESTORING THE WILD

This chapter gives a short account of some of the component parts of the flora of Britain and Ireland, drawing on the excellent resource of detailed botanical publications on the subject. Our present-day flora is the starting point for restoring the wild. The origin of the restoration movement dates back fifty years or so, and the people and events that have played a part are mentioned. Also in outline, I describe restoration species, the different methods of restoration, and briefly consider the range of herbaceous plant communities that may benefit from restoration. This will all be discussed in greater detail in subsequent chapters.

Wild Plants of Britain and Ireland

The wild flora of Britain and Ireland is changing fast. As we enter an interglacial period temperatures are rising, exacerbated by man-made releases of CO_2. Trade and travel are bringing thousands of species from elsewhere to our islands, while intensive land use and development disturb and alter the environment, with many native species having declined alarmingly in abundance over the last 100 years.

The vascular plant species currently present in the British Isles can be divided into those that are recorded in the wild as self-sustaining populations (the wild flora) and those that are cultivated and do not, at present, occur in the wild as self-sustaining populations (the cultivated flora). Native species, archaeophyte species and naturalised neophyte species, together totalling around 3,000 species, constitute the wild flora of the British Isles (Stroh et al., 2023). The cultivated flora consists of 50,000–80,000 non-naturalised neophyte species and cultivars.

Native Species

Currently, our native flora consists of around 1,400 species, rising to around 2,000 species if apomictic 'micro-species' of apomictic genera (such as *Taraxacum*, *Hieracium*, *Rubus*) are also included. Native species are those that were naturally present during the last glaciation and have survived to this day, those that have colonised the British Isles naturally since the last glaciation, and those that have arisen naturally *de novo* since the last glaciation. The tally of native species in the British Isles continues to change through extinction, natural colonisation and speciation, but the total number of native species is relatively low, and moderately constant. The short interval between the last glaciation and the present time has

Yellow Rattle (*Rhinanthus minor*), Common Spotted Orchid (*Dactylorhiza fuchsii*) and Meadow Cranesbill (*Geranium pratense*) growing side by side in a meadow.

provided very little opportunity for the evolution of endemic species. The total number of native species in Britain and Ireland is still rebuilding following mass extinctions during the last Ice Age, but only very slowly due to geographic isolation and the counter pressure of habitat destruction and development.

Archaeophyte Species

This is a group of around 197 species that are non-native but are now well naturalised in the wild; they were introduced to the British Isles intentionally or unintentionally by human activity in the post-glacial period from Neolithic times to AD1489. The list of archaeophytes includes many annual 'weeds' associated with past crop and garden cultivation. Their present and past distributions mirror patterns of historic human habitation and arable cultivation, although

Common Poppy (*Papaver rhoeas*) is an archaeophyte and one of the top five wild species for providing pollinator services.

many annual archaeophytes have contracted alarmingly in range in the twentieth century due to chemical and modernised farming methods.

Naturalised Neophytes

Naturalised neophytes are those species that were intentionally introduced to the British Isles in the period since AD1489, or that were unintentionally introduced during this period as an incidental consequence of human activities. AD1489 is a significant date as it approximately marks the start of seafaring adventure worldwide in search of botanical curiosities that may have ornamental, commercial or scientific value, and the 'discovery' of America. Collection and importation of botanical specimens from across the world continues to this day, and thousands of non-native species have been introduced to the British Isles in this way.

Naturalised neophytes are those introduced alien plant species that have jumped the garden fence, or have escaped horticultural, forestry and agricultural cultivation, or have escaped from botanic gardens, and have established reproducing and self-sustaining populations in the wild since AD1489. Notable examples are Snake's-head Fritillary (*Fritillaria meleagris*), Common Field-speedwell (*Veronica persica*), Oxford Ragwort (*Senecio squalidus*) and Hedgerow Cranesbill (*Geranium pyreniacum*). Rhododendron (*Rhododendron ponticum*) is also a naturalised neophyte in the British Isles, having been first introduced as a garden ornamental around 1763, and then escaping to colonise the wild.

The tally of naturalised neophytes is growing rapidly; it is currently at around 1,500 species, adding significantly to the

Snake's-head Fritillary (*Fritillaria meleagris*) is a naturalised neophyte and has become the iconic wild flower of floodplain meadows.

species diversity of the wild flora of the British Isles. At present, naturalised neophytes are most frequent in urban areas, where they may be the dominant ground cover of the urban wild. These naturalised neophytes are recent alien newcomers to our flora, and are gradually dispersing out from centres of habitation into the wild. Most are a curiosity, some are invasive and a threat to biodiversity, and a small number find a welcoming niche in an existing ecosystem, as has Snake's-head Fritillary on floodplain meadows.

Non-naturalised Neophytes

Non-naturalised neophytes are overwhelmingly the largest grouping of plant species found in the British Isles, existing in cultivation and in collections rather than in the wild. They are those introduced alien species that have so far proven to be unable to reproduce in the wild, and have not so far established self-sustaining populations in the wild, or have not had the opportunity to do so. If ever found in the wild, it is usually as short-lived transient individuals, perhaps as garden throw-outs, and which may be termed 'adventive' or 'casual'. To survive as growing plants in cultivation, these aliens must normally be tended and protected from the rigours and challenges they may face if growing in the wild. Such species are grown for ornament, curiosity and for produce, in gardens and parks, and in horticulture, agriculture and forestry. The Royal Horticultural Society lists over 69,000 'domesticated' neophyte species and cultivars as available from British Nurseries (RHS, 2022). A further 50,000–80,000 non-naturalised neophyte species are held in British botanic gardens as seed *in vitro* in gene banks, or as individual plants growing in pots or in garden beds within the confines of botanic gardens.

Non-naturalised neophytes have no role in nature recovery or in the creation and restoration of wild vegetation in Britain and Ireland, although many will have a role in

Municipal planting using traditional garden plants, most of which are alien non-naturalised neophyte species.

their native place of foreign origin. Similarly, there is no role for cultivars of ornamental and crop species as these have been bred from wild stock, and in the process have lost genetic variation and adaptations to the wild.

Identifying Wild Plant Species

Looking at the wild plants in a meadow or amongst other wild vegetation, it will appear, at first glance, that individual plants are either similar to one another, or dissimilar to one another. Individual plants appear to cluster naturally into groups of similar-looking plants that we instinctively recognise as species. Their distinguishing phenotypic features allow the grouping of similar-looking individuals into species, and the separation of dissimilar-looking species one from another. Species may then be grouped into genera, and genera into families, where the member species of a genus or family may possess a set of common distinguishing features.

Finding an unfamiliar species in the wild sparks curiosity, and putting a name to a new find is sometimes a challenge, but always rewarding. Having a name for a species opens an entry point to the literature on that species. The current standard reference identification flora is *The New Flora of the British Isles* by Clive Stace (Stace, 2019). Prior to 1991, *The Flora of the British Isles* (Clapham et al., 1962) was the standard work. Of course, we also have many British floras where identification can be achieved by leafing through pictures. *The New Concise British Flora* (Keble Martin, 1965) remains the classic.

Having found and named a wild plant, the discovery can be checked against current and historic distribution records by referring to the Botanical Society of Britain and Ireland (BSBI) plant-distribution maps at hectad (10km^2) or tetrad (2km^2) resolution (Stroh et al., 2023). County floras will often have distribution maps at tetrad or monad (1km^2) resolution.

Restoration and Creation

Restoration: The diversification of existing species-poor, semi-improved (agriculturally semi-improved) grasslands by over-seeding with additional species, to restore degraded wild communities.

Creation: The seeding of bare prepared soil with a complete mixture of meadow species, often on former arable land, but also in gardens, parks and in landscaping, to create new wild communities.

Botanic gardens are the Noah's Ark of wild botanical species diversity, safeguarding small representative seed samples of each species *ex situ*/*in vitro* in case of a catastrophic event. Species samples are often stored in 1oz medicine bottles or Kilner jars, held at low moisture

Three reference books that may be used to identify wild plants and check on distribution.

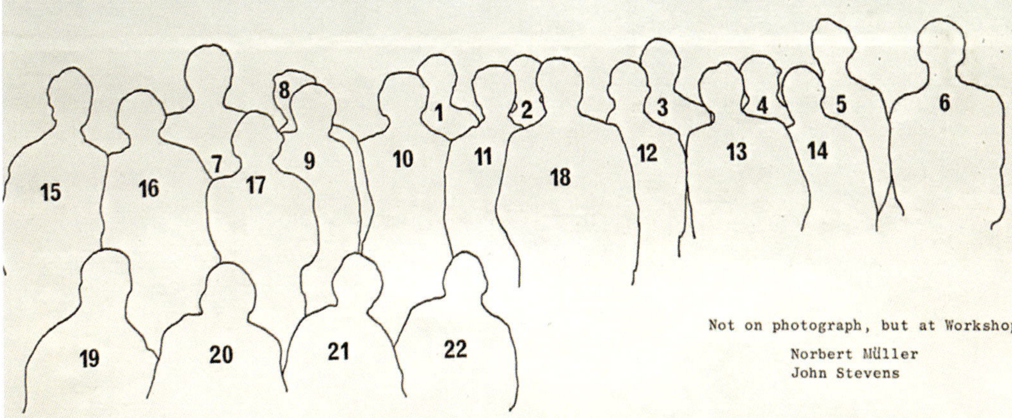

Not on photograph, but at Workshop:

Norbert Müller
John Stevens

1	Johnathon Franks	2	Heni Koningen	3	Wolfgang Schumacher	4	Alan Frost
5	Brian Davis	6	Donald MacIntyre	7	Rainer Kretz	8	Derek Wells
9	Wolfram Kunick	10	Chris Baines	11	Andreas Winkler	12	Geoff Taylor
13	Marten Håmmer	14	Max Hooper	15	'Loki' Schmidt	16	Miriam Rothschild
17	Lore Lässer	18	Terry Wells	19	Alex Oberholzer	20	John Chambers
21	Rob Leopold	22	Michael Avishai				

Participants at a conference on meadow creation held at Monks Wood Experimental Station, Huntingdon, in 1983. Organised by Dr Terry Wells, it was the first such meeting on the subject and attracted participants from across Europe.

content and low temperature. Botanic gardens cannot realistically be expected to store and safeguard the huge and important biodiversity represented by 'within species' genetic diversity (Volis and Blecher, 2010). This cannot be done in a bottle. The only way to conserve *within species* biodiversity is to protect and nurture this biodiversity in existing or restored healthy ecosystems, *in situ/in vivo*.

With foresight, in the mid-1970s, the Nature Conservancy Council (NCC) recognised the need to reverse declines in biodiversity, and judged that meadow creation could play a part in that recovery. Research was commissioned by Dr Derek Wells of the NCC and led by Dr Terry Wells of The Institute for Terrestrial Ecology (ITE) at Monks Wood Experimental Station, Huntingdon.

This research comprised experimental trial plots producing seed of a range of wild species and other trial plots testing different combinations of seed of these species when sown together. I recall visiting Terry Wells in 1981, and walking the plots with him. He gave me a box of 35mm film canisters, each containing stock seed of a range of different wild species. These stocks formed the foundation of multiplication and production plots grown over the following years by others and myself. Terry Wells and his team published methodologies, most notably 'Creating attractive grasslands using native plant species' in 1981 and later (Wells *et al* 1981, 1986, 1989).

The pioneering work of Dr Terry Wells and colleagues created great interest in the British Isles and across Europe at that time. In 1983 the first 'Wild Seeds Workshop' was held at Monks Wood Experimental Station, bringing together many people from across Europe who had been independently thinking and working along the same lines. The participants are shown in the photo on the previous page, and included Dr Terry Wells, Dr Derek Wells, Chris Baines, Loki Schmidt (wife of the former German Chancellor), Dame Miriam Rothschild, Rainer Kretz, Max Hooper, John Chambers, Rob Leopold, Michael Avishai, myself, and others. In addition to Terry and Derek Wells, Miriam Rothschild deserves a special mention for the encouragement she gave me, and the work she did to promote meadow creation and restoration in those early days.

At that first meeting in 1983, and over the intervening forty years to the present, there has been growing enthusiasm and increasing acceptance of the benefit and need to create and restore species-rich meadows and other wild communities. The concept was not at first completely welcomed by some in the conservation world, who thought that sowing wild seeds would 'mess up the countryside'.

In 1984 I was approached by the then Nature Conservancy Council (now Natural England) and offered a plot on North Meadow National Nature Reserve (NNR) for seed harvest the following year. That presented an exciting challenge, because at that time there was no dedicated method for harvesting meadow seeds. There followed trials of a series of different and improving prototype harvesters made by myself, culminating in the making of the first meadow brush harvester in 1986, and the first successful harvest of meadow seeds from North Meadow NNR by that means that year.

Brush harvesting and restoration using meadow seeds was soon adopted by many County Wildlife Trusts to repair and expand their reserves, and has also been embraced by local authorities, utilities and business, landscape architects and farming, and more recently, gardening and wilding, all contributing to the restoration of nature and the recovery of ecosystems.

Today there is a place and a role for meadow restoration and creation almost anywhere, including windowbox and garden, road and rail networks, parks and open space, farmland and forestry. And each year, wild seed production across the world grows, as does the stock of restored and created meadows and habitats worldwide. But there is a long road ahead and much more to be done.

Around the world the need to restore nature has driven research on the fundamental challenges facing ecological restoration, it has inspired the formation of structures and institutions to assist and regulate native seed production, and it has generated finance to restore native ecosystems on degraded land, in the United States (Oldfield et al., 2019), Germany (Mainz and Wieden, 2018), France (Rivière et al., 2022), Australia (Hancock et al., 2020) and many other countries.

The Vegetation Types Covered in this Book

In this book we look at the use of non-woody terrestrial vascular plant species (mostly native) that may be used to restore and create herbaceous vegetation and ecosystems. We include those species that we would normally recognise as ferns, wild flowers, grasses, sedges and rushes. For convenience we exclude mosses, liverworts, lichens and other lower green plants as these are not easily amenable to restoration and are, to a moderate extent, natural colonists of suitable restoration sites. Together, green plants are important because they are

Bird's-foot Trefoil (*Lotus corniculatus*), an important constituent species of native wild grasslands.

the primary producers in ecosystems, harnessing photosynthesis to build the biomass, energy and structure on which all other terrestrial life forms depend. Green plants ultimately feed the bacteria, the fungi, the molluscs, the nematodes, the invertebrates (insects), the amphibia and reptiles, the birds and the mammals that together make up a healthy wild ecosystem. There will be situations where some of these creatures may need assistance in colonising restored ecosystems (for example Large Blue butterfly, mega fauna), but none will survive without healthy and biodiverse wild vegetation to call home.

The vegetation of the British Isles varies from location to location, resulting in different recognisable types of vegetation and plant communities. These different types of vegetation are excellently described for the British Isles in the *National Vegetation Classification* (NVC), (Rodwell, 2006). Published in five volumes, the NVC classification describes 286 different community types and 578 sub-community types, all grouped into twelve major vegetation types.

Projects, large and small, will often seek to restore and create a diversity of habitats. These different habitats may be chosen and combined in a way that respects the land-use history of the site, the existing vegetation and features on the site, while also fulfilling the wishes of the landowner or land manager. Meadows are important, but so are other habitats, and when combined they can, together, provide greater benefit to nature than in isolation.

For convenience, the 286 NVC community types and 578 NVC sub-community types are treated as eleven generalised vegetation types in the pages that follow. This allows a more general approach to describing principles and methods of wild herbaceous vegetation restoration and creation. However, the NVC remains the standard and most accurate classification to describe existing vegetation, and may be used as a guide (but not as a prescription) on which to base detailed objectives and species composition for wild vegetation restoration and creation.

GENERALISED VEGETATION TYPES

- Annual wild flowers
- Neutral lowland meadow and pasture
- Calcareous meadow and pasture
- Vegetation on sandy soils
- Floodplain, wetland and water edge
- Upland meadow and pasture
- Moor and heath
- Wayside and tussock vegetation
- Hedgerow and scrub
- Woodland ground flora
- Coastal

CHAPTER 2

BIODIVERSITY, SPECIES, ECOTYPES AND ECOSYSTEMS

This chapter looks at the astonishing diversity of life in meadows and other wild herbaceous vegetation and discusses how individual wild plants cluster into distinctive local ecotypes, how ecotypes cluster into distinctive species, how species cluster into communities, and how communities build ecosystems.

Biodiversity

Biodiversity is the sum of the genetic variation of all the living organisms in a unit of volume. Take, for example, a cubic metre of meadow. With meadows, as with most habitats, there is far more living biomass and more genetic diversity below ground than there is above ground. The genetic diversity between and within all life forms, above and below ground, is the basic and fundamental measure of biodiversity. By this measure, a cubic metre of meadow is more biodiverse than the contents of an equivalent volume of the deep freeze of a gene bank.

We can measure and record the genetic code, but we cannot see and read it with our naked eyes. We can, however, see the diversity of species in a meadow, and the variation within those species. The Convention on Biological Diversity, Rio de Janeiro (1993), defines biodiversity as follows:

> *Every living thing, including the natural patterns they form. Including every species and all the genetic differences within each species. The variety of ecosystems: forests, drylands, wetlands, mountains, lakes, rivers, agricultural lands, seas and islands where living*

creatures, including humans, animals, insects and plants, that form a community, interacting with one another, and with the air, water and soil around them.

High genetic variation (biodiversity) across species, between ecotypes within species, between individuals within ecotypes, and within individuals, is important in grassland restoration and creation because it creates genetically robust communities that are best able to survive and respond to environmental heterogeneity across space and time.

Common Broomrape (*Orobanche minor*) and Bee Orchid (*Ophrys apifera*), delightful examples of the diversity of our native flora that may occasionally be found in surviving wild places.

Species, Ecotypes and Communities

Species

A species is a natural grouping of individuals that cluster and share with each other a common set of defining features that are not shared with other species. These salient defining features may be the visible morphology (the macroscopic and microscopic phenotype) of the species, or unique information held in DNA base sequences (the genotype), or a combination of both. Each defining feature of a species will have a genetic mean and a genetic variance. The mean is the average measure of that feature for the species, and the genetic variance is the spread of the genetic measure of that feature around the mean. The genetic variance of the defining features of two closely related species must not overlap, otherwise they are the same species. For a more critical account of species concepts *see* Barraclough and Humphreys (2014).

The following page presents a schematic representation of the genetic mean and variance of two related species that differ in height. Plotting frequency against feature size will often produce a symmetrical or skewed normal curve.

The individuals of a species usually share a common gene pool, and experience few barriers to interbreeding with each other, except that individuals of outbreeding species are not normally able to self-fertilize, and individuals of selfing species are not normally able to cross with other individuals of the same species.

We are indebted to science for our understanding of the origin of species, and in particular to Charles Darwin

Hedge Bedstraw (*Galium album*), Bird's-foot Trefoil (*Lotus corniculatus*) and Yellow Rattle (*Rhinanthus minor*) growing together in the wild.

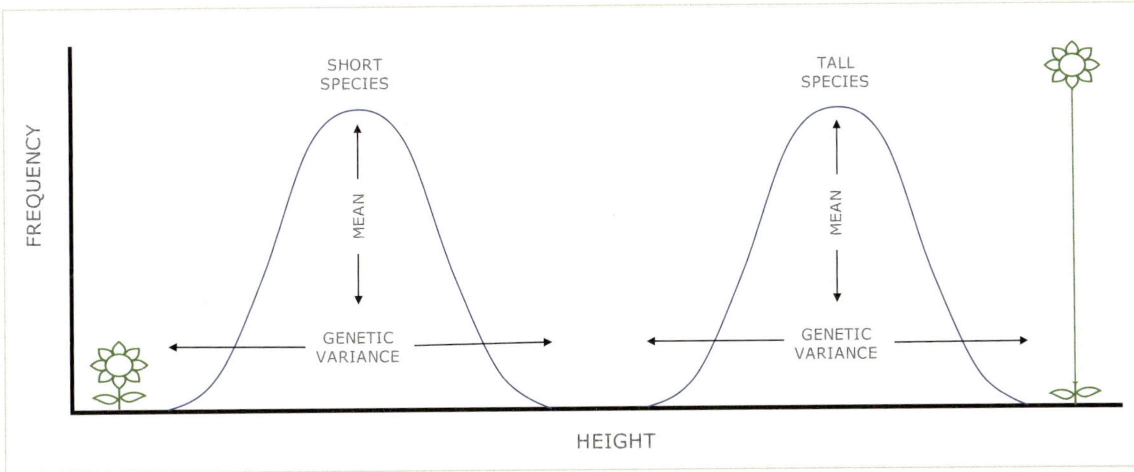

Figure 1: Schematic normal distribution showing genetic mean and genetic variance for height of hypothetical short and tall species. Note that the genetic variance for height of the two species does not overlap.

Flower spikes of Common Spotted Orchid (*Dactylorhiza fuchsii*) in clonal pairs from a single population exhibiting considerable genetic variation between pairs.

(1859) for 'evolution by natural selection', to Gregor Mendel (1866) for 'the mechanism of inheritance', and to Watson and Crick (1953) for 'the structure of DNA'.

Not all species in the plant world are equal. At one extreme are those species that are largely asexual, reproducing by vegetative means or by apomixis. These asexual species tend to be uniform, any existing heterozygosity is locked in, and any evolution is slow and step wise, mediated by somatic mutation (or other mechanisms as yet unknown). This mode of reproduction can produce arrays of distinct clones. Many of these have been named, and are referred to as 'micro-species'. The origin of micro-species is uncertain. They may be simple 'sports', or they may be more complex accumulations of somatic mutations in response to natural selection by a process akin to speciation, over very long periods of time. There are around 260 micro-species of Dandelion (*Taraxacum officinale*), 100 or so micro-species of Wild Rose (*Rosa canina*), and 200 or so micro-species of Hawkweeds (*Hieracium sp*).

At the opposite extreme are those species that reproduce sexually, are highly fertile, and have weak barriers to hybridization between related species. The result is that such species are very variable, may exist as complex hybrid swarms, and are often actively evolving. As may be expected, it can be difficult to define the limits of these species with any precision. A good example of a dynamic evolving genus is *Dactylorhiza* (the Spotted Orchids). Species of *Dactylorhiza* readily hybridise frequently with each other, and genetic variation within species is huge. Interestingly, this genetic variation is nicely demonstrated in the wild in hybrid swarms involving Common Spotted Orchid and Southern Marsh Orchid. *Dactylorhiza* is polycarpic and may produce clonal flower spikes side by side following division of the tubers. These clonal pairs enable comparison with other paired individuals nearby, thus demonstrating the environmental influences on the phenotype, and revealing by observation the astonishing diversity of genetic variation in *Dactylorhiza*. The images on the previous page provide a few examples taken from a single population.

Ecotypes

For simplicity, we group together, as *ecotypes*, all those below-species ranks sometimes used in classification (such as sub-species, variety, sub-variety, form), while retaining *cultivar* for the cultivated forms that are derived from the artificial breeding of wild species.

An ecotype is a distinct population within a naturally variable wild species, and ecotypes are a natural selection response to climatic, edaphic (nature of the substrate) and biotic (interactions with other species) factors, or to isolation. Different ecotypes within a species may be distinguished one from another by defining morphological or genetic features, and often a defining place of origin.

Species and ecotypes are different. Like species, individual ecotypes may be confined to one location, or may be

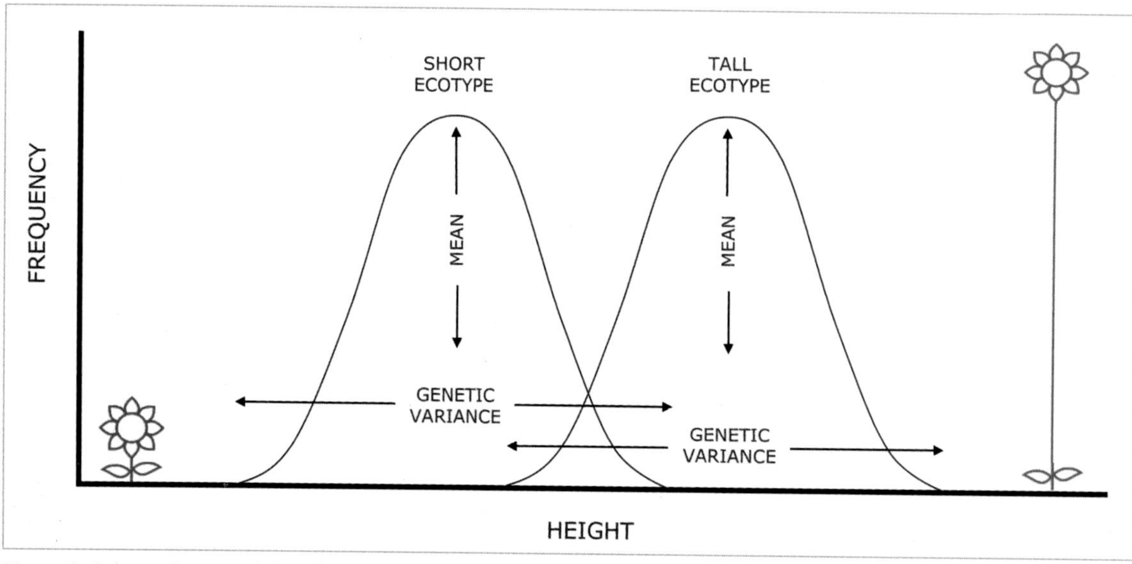

Figure 2: Schematic normal distribution showing genetic mean and genetic variance for height of hypothetical short and tall ecotypes. Note that the genetic variance for height of the two ecotypes overlaps.

found at different locations. However, the defining features of a species will have a genetic variance that does not overlap with that of related species (otherwise they would be the same species). The salient defining features of an ecotype have means that are distinct from those of other ecotypes of the same species, but the variance of the defining features usually overlaps with that of other ecotypes of the same species (otherwise the ecotype would be a species).

Figure 2 on the previous page is a schematic representation of ecotypes with frequency plotted against feature size, in this case inflorescence height. The frequency distribution of short and tall ecotypes takes the form of two overlapping normal curves with different means.

Wild ecotypes are frequent throughout the wild British and Irish flora. They are a huge reservoir of living biodiversity, essential for the proper functioning of ecosystems. They convey a local sense of place, they have heritage value, their scientific and utilitarian value is priceless, and they are a living genetic response to a changing environment, in both space and time. Therefore ecotypes are an important ecosystem component of restored, created and established species-rich meadows and other herbaceous vegetation, as is the dynamic evolutionary process that generates and sustains ecotypes.

Communities

Communities are assemblages of individuals of different species. The plant-to-plant and species-to-species interactions within a community are critical for determining the structure and the dynamics of the community (Bilas et al., 2020). Those interactions may be by *wired* communications between plants mediated by mycorrhizal hyphae, parasite/symbiont/host interactions, or *wireless* communication mediated by volatile emissions and root exudates (Sharifi and Ryu, 2020), as well as direct and indirect physical, chemical and radiological contact between plants (Smith et al., 2020, and Veits et al., 2018). Community species richness and structural diversity increases over time under favourable conditions, and this increases ecosystem productivity and stability.

Ecosystems

An ecosystem is a natural living entity that consists of organisms interacting one with another through intimate (Anten and Chen, 2021) and distant web-like contacts (Adamatzky, 2022), and also interacting with the environment around them, through space and time. The spatial limits of an ecosystem are defined by those boundaries across which there is no interaction.

Clearly, an ecosystem may be as small as a pond, or as large and complex as a forest. As the complexity and size of an ecosystem increases, so does its value to life on earth. There is a positive relationship between species diversity and ecosystem structural complexity (Coverdale and Davies, 2023), and this drives ecosystem productivity, with ecosystem services output increasing with increasing species diversity, at least to a point at which saturation is approached. Species population size and the spatial distribution of individuals of a species within an ecosystem also influence ecosystem productivity, with more even population size and species distribution across an ecosystem yielding greater ecosystem service productivity than otherwise (Hordijk et al., 2023).

CHAPTER 3

PROTECTED, SENSITIVE AND PROSCRIBED SPECIES

In this chapter we look at those native species of the flora of Britain and Ireland that may not, under normal circumstances, be used in the restoration and the creation of wild vegetation. These native species are those that are 'protected' by law, those that are 'sensitive' for scientific or heritage reasons, and those that are 'proscribed' under Seeds Regulations.

Protected Species

The Wildlife and Countryside Act 1981

UK Public General Acts: 1981 *c*.69

The Wildlife and Countryside Act 1981 lists, on Schedule 8, those species of wild plant (*c*.69) that have the greatest level of protection in England, Wales and Scotland. It is illegal to destroy, harm, pick, possess, trade or advertise for trade, any part of a wild protected species listed on Schedule 8 of the Act. To do so is an offence subject to a maximum penalty of six months in prison.

Most species on Schedule 8 are extremely rare in the wild and their populations are often small and vulnerable. Many of these species are obscure and unlikely to be encountered, while others may be familiar. The feature box lists some of the better known terrestrial species.

Bluebell (*Hyacinthoides non-scripta*) is on Schedule 8, but only so far as it is illegal to offer for sale wild collected seed or bulbs of Bluebell. Wild collection of seed is permitted under licence from Natural England, and seed and plants produced in cultivation are exempt. Bluebell was added to Schedule 8 in 1998 following reports of extensive digging of wild Bluebell bulbs in woods in Norfolk, North Wales and elsewhere, for sale 'in the green'. This is an illustrative example of wildlife crime in the British Isles, and there have been a number of successful convictions under the provisions of the 1981 Act for digging wild Bluebells.

Some of the protected species on Schedule 8 are in horticultural cultivation and are freely available as *garden forms*, for example Meadow Clary (*Salvia pratensis*). It remains unlawful to possess wild origin material of Meadow Clary, and therefore this and other protected species of wild origin cannot be used in restoration projects, unless done so under licence from Natural England. A consequence is that it is perfectly legal to plant the garden form of Meadow Clary in the wild, but not the true wild form of the species. Planting or sowing any garden cultivar in the wild is potentially harmful to native species biodiversity.

Additional Protection for Wild Plants

All non-Schedule 8 wild plants in the British Isles have some additional basic legal protection. Under Section 13 of The Wildlife and Countryside Act 1981, it is an offence to uproot any wild plant without the permission of the landowner or the land occupier, on both public and private land. Of course, the landowner/occupier does not require authority to uproot a wild plant on their own land, unless it is a Schedule 8 species. The definition of 'uprooting' was tested in a case centred on the collection of *Sphagnum* moss from common land, again in Norfolk, where it was ruled that the pulling of moss amounted to uprooting.

The collection of seed from the wild is less clearly defined. Entering private land without permission is trespass, a civil matter, and the collection of seed from private land is an offence only if it can be shown that doing so causes criminal damage. Entering public land is not trespass, and picking plant material (such as blackberries) on public land is not an offence, unless doing so causes criminal damage, or if doing so is prohibited by byelaws. Some National Trust land with public access is, for example, covered by byelaws prohibiting the collection of plant material.

Sensitive Species

'Sensitive species' are those species that, for reasons of science, phytogeography and heritage, are best precluded from being used in ecosystem restoration, except in very special circumstances. Precluded are those species, microspecies and distinctive ecotypes that are rare, have a local or restricted distribution in the British Isles, have a long history of a restricted pattern of distribution, or exist in the wild as distinct isolated populations.

The native British flora consists of around 1,400 species, of which around 1,000 may be considered to be sensitive. Examples of sensitive species (that may be familiar) are given in the feature box. At present, a complete definitive list of sensitive species does not exist for the British Isles.

PROTECTED SPECIES

Examples of some of the fully protected terrestrial plant species listed on Schedule 8 of The Wildlife and Countryside Act 1981

Alyssum alyssoides (Small Alison) NN
Cephalanthera rubra (Red Helleborine) N
Cypripedium calceolus (Lady's-Slipper Orchid) N
Dianthus armeria (Deptford Pink) N
Dianthus gratianopolitanus (Cheddar Pink) N
Epipogium aphyllum (Ghost Orchid) N
Gagea serotina (Snowdon Lily) N
Gentiana nivalis (Alpine Gentian) N
Gentiana verna (Spring Gentian) N
Gladiolus illyricus (Wild Gladiolus) N
Himantoglossum hircinum (Lizard Orchid) N
Hyacinthoides non-scripta (Bluebell) N
Liparis loeselii (Fen Orchid) N
Lythrum hyssopifolia (Grass-Poly) N
Melampyrum arvense (Field Cow-Wheat) NN
Mentha pulegium (Pennyroyal) N
Orchis militaris (Military Orchid) N
Orchis simia (Monkey Orchid) N
Ophrys sphegodes (Early Spider-Orchid) N
Orobanche caryophyllacea (Bedstraw Broomrape) N
Orobanche picridis (Oxtongue Broomrape) N
Orobanche reticulata (Thistle Broomrape) N
Polygonatum verticillatum (Whorled Solomon's-Seal) N
Rhinanthus angustifolius (Greater Yellow-Rattle) N
Salvia pratensis (Meadow Clary) N
Senecio paludosus (Fen Ragwort) N
Silene suecica (Alpine Catchfly) N
Veronica spicata (Spiked Speedwell) N

N = Native, A = Archaeophyte,
NN = Naturalised Neophyte

OPPOSITE: Lizard Orchid (*Himantoglossum hircinum*), a very rare protected species listed on Schedule 8 of the Wildlife and Countryside Act 1981.

SENSITIVE SPECIES

Examples of some native terrestrial plant species (not on Schedule 8) that would not normally be considered candidates for use in ecosystem restoration projects due to their scientific and heritage sensitivity.

Aceras anthropophorum (Man Orchid) N
Ajuga chamaepitys (Ground-Pine) N
Alchemilla, local apomictic micro-species (Lady's Mantles) N
Armeria maritima ssp elongata (Thrift) N
Botrychium lunaria (Moonwort) N
Cephalanthera damasonium (White Helleborine) N
Cephalanthera longifolia (Sword-Leaved Helleborine) N
Coeloglossum viride (Frog Orchid) N
Corallorhiza trifida (Coralroot Orchid) N
Dianthus deltoides (Maiden Pink) N
Dryas octopetala (Mountain Avens) N
Gentiana pneumonanthe (Marsh Gentian) N
Gentianella germanica (Chiltern Gentian) N
Hammarbya paludosa (Bog Orchid) N
Helianthemum apenninum (White Rock-Rose) N
Herminium monorchis (Musk Orchid) N
Hieracium, local apomictic micro-species (Hawkweeds) N
Neotinea ustulata (Burnt Orchid) N
Neottia cordata (Lesser Twayblade) N
Neottia nidus-avis (Bird's-Nest Orchid) N
Ophioglossum azoricum (Small Adder's Tongue) N
Ophioglossum lusitanicum (Least Adder's Tongue) N
Ophrys insectifera (Fly Orchid) N
Osmunda regalis (Royal Fern) N
Polemonium caeruleum (Jacob's Ladder) N
Primula elatior (Oxlip) N
Primula farinosa (Bird's-Eye Primrose) N
Primula scotica (Scottish Primrose) N
Pulsatilla vulgaris (Pasque Flower) N
Rhodiola rosea (Roseroot) N
Taraxacum, local apomictic micro-species (Dandelions) N

N = Native, A = Archaeophyte,
NN = Naturalised Neophyte

Proscribed Species

Proscribed species are those species listed in domestic legislation governing the marketing of seeds and plant propagation material. The legislation covers those plant species that are important as agricultural, horticultural and forestry crop species. The purpose of the legislation is to guarantee minimum standards of purity, germinability and trueness to type, at the point of sale, and to provide a structure allowing for the collection of royalty payments by the breeder. This is of interest to us because some of the proscribed crop species are also native wild species of the British Isles.

The Vegetable Seeds Regulations 1993

UK Statutory Instruments: 1993 No. 2008

These regulations restrict the marketing of seeds of the vegetable species listed in Schedule 1 of the regulations. For these vegetable species only those cultivars that are listed in the 'Common Catalogue/National List' may be marketed. The only species that may concern us here is Wild Carrot (*Daucus carota*), as this species is listed in Schedule 1 of the regulations. Of course, *Daucus carota* is a native wild species widespread in the British Isles. Restoration of wild populations of this species, to be compliant with the Vegetable Seeds Regulations 1993, requires using seed only of Common Catalogue/National List cultivars of the garden carrot.

The regulations also create a problem for those who wish to maintain and grow land races and old cultivars of the vegetables. The marketing of land races of vegetables is prohibited by these regulations.

The Marketing of Ornamental Plant Propagating Material Regulations 1999

UK Statutory Instruments: 1999 No.1801

These regulations apply minimum standards to the marketing of seed and live plant material of ornamental species for vigour, germinability, purity, identity and freedom from disease. This legislation does not define an ornamental species or provide a schedule of the species to which the legislation applies, but it does exempt from the legislation 'the marketing of material for use in the conservation of genetic diversity'.

The Seed Marketing Regulations 2011

UK Statutory Instruments: 2011 No.463
(EU Seeds Directive 2010/60/EU)

The Seed Marketing Regulations 2011 (as amended) currently govern the marketing of agricultural and horticultural seeds in England. Under the regulations, only certified seed of Common Catalogue/National List cultivars of the proscribed species in Schedule 1 may be marketed lawfully. Of the fodder species listed in Schedule 1, twenty-three are also wild native species of the British Isles, and most of these species are important constituent species of wild meadows, pastures and other grasslands. This is unfortunate for us because the regulations do not permit the marketing of the wild form of any of the twenty-three proscribed native species.

However, the regulations do permit the marketing of wild harvested mixtures containing (unavoidably) uncertified wild seed of proscribed species, provided a pink label is attached describing the seed as a 'Preservation Mixture', and provided various regulatory, compliance and record-keeping criteria are adhered to. The author was party to pre-legislation consultation meetings on this matter, and is delighted that his faded pink polo inspired the choice of colour for the Preservation Mixture label.

The 'Preservation Mixture' pink label is a small concession, by plant breeders, allowing limited marketing of wild seed of the twenty-three proscribed native species in wild harvested mixtures only. Unfortunately, no legislation exists in the British Isles to specifically allow, support, control and apply standards to the marketing of wild seeds of native species, both proscribed and others (other than The Trades Description Act 1968). It is therefore inevitable that wild seed of proscribed species, such as Bird's-foot Trefoil (*Lotus corniculatus*), Red Clover (*Trifolium pratense*), Common Vetch (*Vicia sativa*) and Sheep's Fescue (*Festuca ovina*), will be produced and marketed for nature restoration purposes, outside the legal framework of the Seed Marketing Regulations 2011.

PROSCRIBED SPECIES

Proscribed 'fodder' species listed in Schedule 1 of the Seed Marketing Regulations 2011

Agrostis canina (Velvet Bent) N
Agrostis capillaris (Brown Top) N
Agrostis stolonifera (Creeping Bent) N
Arrhenatherum elatius (False Oat Grass) N
Cichorium intybus (Chichory) A
Dactylis glomerata (Cocksfoot) N
Daucus carota (Wild Carrot) N
Festuca ovina (Sheep's Fescue) N
Festuca rubra (Red Fescue) N
Lolium perenne (Perennial Ryegrass) N
Lotus corniculatus (Bird's-Foot Trefoil) N
Medicago lupulina (Black Medick) N
Onobrychis viciifolia (Sainfoin) (N)
Phleum bertolonii (Small Timothy) N
Phleum pratense (Timothy) N
Poa annua (Annual Meadow-Grass) N
Poa nemoralis (Wood Meadow Grass) N
Poa pratensis (Smooth-Stalked Meadow Grass) N
Poa trivialis (Rough-Stalked Meadow Grass) N
Schedonorus arundinaceus (Tall Fescue) N
Schedonorus pratensis (Meadow Fescue) N
Trifolium hybridum (Alsike Clover) NN
Trifolium pratense (Red Clover) N
Trifolium repens (White Clover) N
Vicia sativa (Common Vetch) N

N = Native, A = Archaeophyte, NN = Naturalised Neophyte

CHAPTER 4

INVASIVE ALIEN WEEDS AND NATIVE WEEDS

This chapter is all about weeds, both invasive alien weeds and native weeds. So what is a weed, and why should we not want to use weed species in restoration seeding? The answer is that a weed is a species that is subjectively or objectively perceived to do more harm than good (Merfield, 2022). Such harm may include:

- Harming biodiversity and the environment
- Injurious to us due to toxicity or physical harm
- Injurious to our livestock or our pets, due to toxicity or physical harm
- Competitive in the field and garden resulting in harm to our cropping and cultivation
- Causing structural harm to our buildings
- Visually harming the aesthetic of our surroundings.

Weed species, as defined here, are generally excluded from use in ecosystem restoration and creation.

Invasive Alien Weed Species

Invasiveness is the ability to colonise new territory quickly, sometimes ousting native species and causing harm. Moderate invasiveness is not necessarily a harmful feature of a species. Of course, our native flora is invasive by definition, otherwise our native species would not be here, having, of necessity, invaded our islands by natural means following the last Ice Age. By contrast, harmfully invasive alien species of flora and fauna are a big problem. Alien species have been transported across the globe by humans for centuries, both intentionally and unintentionally, and some of these introduced alien species have seriously harmed native ecosystems, particularly those ecosystems that were formerly quite isolated (Turbelin et al., 2017).

Cheat Grass (*Anisantha* (*Bromus*) *tectorum*) is a good illustrative example of a harmful invasive alien species. Cheat Grass is now dominant in many parts of the arid Western USA, especially the Great Basin, there having replaced native Sagebrush ecosystems. Early European settlers with their sheep and cattle over-grazed the species-diverse native Sagebrush (*Artemisia tridentata*) steppe and shrub-land vegetation down to bare soil over this vast arid landscape. Rain came, soils eroded, and bare land provided an opportunity for Cheat Grass. Inadvertently introduced by the settlers, Cheat Grass is a fast growing, competitive winter annual producing copious seeds that readily disperse across the landscape. In summer it dies back and the dry vegetation ignites easily. As an annual it requires exposed soil to establish. Initially, the over-grazing of Sagebrush created these conditions. Continued dominance of Cheat Grass is now favoured by wild fire, all exacerbated by global warming. Wild fires more frequent than once in nine years sustain the Cheat Grass, and prevent the re-establishment of native Sagebrush closed communities. Restoration of the Sagebrush ecosystem using native seeds is seen as the only way to control Cheat Grass (Davies and Leger, 2021).

The British Isles are blessed with a temperate oceanic climate. Our forests do not burn and wild fires on open ground are relatively infrequent. Therefore the opportunities for non-native invasive annual species are few, other than within arable cropping systems. In the British Isles, the threat to ecosystems and native vegetation from introduced invasive alien terrestrial species, potential and/or real, is mostly from perennial naturalised neophytes. Two species have caused more harm than others: Rhododendron (*Rhododendron ponticum*) and Wall Cotoneaster (*Cotoneaster horizontalis*).

Rhododendron is listed on Schedule 9 of The Wildlife and Countryside Act 1981 (amended 2022), and is classed as a harmful and invasive alien species. However, pollen records in peat and lake sediments reveal that Rhododendron was a native of the British Isles in some interglacial periods prior to the current interglacial period, but is an invasive alien at present. It is now a widespread and naturalised neophyte, a garden escapee, first being introduced to the British Isles as an ornamental species in 1763 (CABI, 2019). The leaves are toxic to livestock, the nectar in the flowers is poisonous to some bee species (Egan et al., 2016), and Rhododendron has spread to become the most abundant and most harmful non-native invasive species in the British Isles, and in doing so has displaced native ecosystems and harmed biodiversity.

Wall Cotoneaster is also listed on Schedule 9 of The Wildlife and Countryside Act 1981 (amended 2022). Introduced to the British Isles in 1824 as an ornamental, it is another garden escapee. Now widespread in the wild as a naturalised neophyte, it behaves as an invasive alien species and out-competes native vegetation on limestone pavement, cliffs and scree, harming biodiversity at a local scale. In less sensitive locations, species of *Cotoneaster* can

Wall Cotoneaster (*Cotoneaster horizontalis*), an example of a harmful invasive alien species listed on Schedule 9 of the Wildlife and Countryside Act 1981.

sometimes support biodiversity by providing good pollinator services and autumn berries, but it is very difficult to control once it starts to spread invasively in the wild.

The Wildlife and Countryside Act 1981 (amended 2022) lists in Schedule 9 the fifty-eight terrestrial and aquatic invasive alien plant species to which the Act applies. Based on current best evidence, these are the species that have the greatest potential to become invasive, or are invasive already, and pose an existing or potential threat of harm to biodiversity in the wild in the British Isles.

It is an offence under the Wildlife and Countryside Act 1981 to intentionally import, keep, breed, transport, sell, grow, cultivate or permit to reproduce, any of the fifty-eight species of listed invasive alien plant species on Schedule 9. This applies to all live parts of these species, including seeds. It also applies to all hybrids, varieties and cultivars of these species. The list of species, and the conditions that apply, are updated every four years. The most serious offence is committed if any of these species, or their derivatives, are intentionally cultivated. The maximum penalty for committing an indictable offence under the Act is two years in prison.

We should remember that many of the species that we now consider to be invasive aliens of the British Isles were collected from around the globe and brought to these islands for reasons of curiosity, commercial potential and ornament for our gardens. Equally, when we have colonised far-off lands we have taken with us land races of our crop species, our native fodder species, and their associated weeds and pests, to the extent that some of the worst invasive alien species of other bio-geographic regions across the world originated from our own native flora. Our native Common Knapweed (*Centaurea nigra*) is a serious invasive alien in North America, and our native Perforate St John's-wort (*Hypericum perforatum*) is an invasive alien in Australia.

In the British Isles there are many naturalised alien species that are only moderately invasive, and occur in the wild as casual intruders. Most have originated as garden escapees or throw-outs, and they can be locally abundant in suitable wild habitat. They may, however, cause subtle harm at a local scale by disrupting evolved mutualistic associations between native species within communities, particularly those complex native plant-pollinator networks and interactions (Parra-Tabla and Arceo-Gomez, 2021).

Not all invasive aliens are unwelcome. Some alien species have completed the citizenship pathway with honours. Snake's-head Fritillary (*Fritillaria meleagris*), first recorded in the wild in 1737 as a probable garden escapee, is now a well naturalised neophyte, and has become the iconic prototypical wild species of floodplain meadows (Walker, 2021). North Meadow at Cricklade, probably the finest lowland meadow in the British Isles, is home to 80 per cent of the wild population of Snake's-head Fritillary and is a National Nature Reserve, partly for this reason.

PROHIBITED INVASIVE ALIEN SPECIES

Prohibited terrestrial invasive alien species listed in Schedule 9 of The Wildlife and Countryside Act 1981. These are the alien species that are currently most seriously invasive, or have the potential to become so.

Allium paradoxum (Few-Flowered Leek) NN
Allium triquetrum (Three-Cornered Leek) NN
Carpobrotus edulis (Hottentot Fig) NN
Cotoneaster bullatus (Hollyberry Cotoneaster) NN
Cotoneaster integrifolius (Entire-Leaved Cotoneaster) NN
Cotoneaster horizontalis (Wall Cotoneaster) NN
Cotoneaster microphyllus (Small-Leaved Cotoneaster) NN
Cotoneaster simonsii (Himalayan Cotoneaster) NN
Crocosmia x crocosmiiflora (Montbretia) NN
Disphyma crassifolium (Purple Dewplant) NN
Gunnera tinctoria (Giant Rhubarb) NN
Gunnera x criptica (Giant Rhubarb) NN
Heracleum mantegazzianum (Giant Hogweed) NN
Impatiens glandulifera (Himalayan Balsam) NN
Lamiastrum galeobdolon subsp argentatum (Variegated Yellow Archangel) NN
Parthenocissus inserta (False Virginia Creeper) NN
Parthenocissus quinquefolia (Virginia Creeper) NN
Reynoutria japonica (Japanese Knotweed) NN
Reynoutria sachalinensis (Giant Knotweed) NN
Rhododendron luteum (Yellow Azalea) NN
Rhododendron ponticum (Rhododendron) NN
Rosa rugosa (Japanese Rose) NN
Smyrnium perfoliatum (Perfoliate Alexanders) NN

N = Native, A = Archaeophyte,
NN = Naturalised Neophyte

Great Quaking-grass (*Briza maxima*), an example of a naturalised neophyte that may sometimes be moderately invasive.

NATURALISED NEOPHYTES

Examples of naturalised alien species (not on Schedule 9) that are sometimes moderately invasive in the wild, or have the potential to become so.

Buddleja davidii (Buddleja) NN
Briza maxima (Great Quaking Grass) NN
Campanula rapunculoides (Creeping Bellflower) NN
Crocus vernus (Spring Crocus) NN
Epimedium alpinum (Barrenwort) NN
Eranthis hyemalis (Winter Aconite) NN
Erythranthe guttatus (Monkey Flower) NN
Erythranthe lutea (Blood-Drop-Emlets) NN
Erythranthe moschata (Musk) NN
Fritillaria meleagris (Snake's-Head Fritillary) NN
Hesperis matronalis (Dames Violet) NN
Hyacinthoides hispanica (Spanish Bluebell) NN
Hyacinthoides italica (Italian Bluebell) NN
Hypericum calycinum (Rose-of-Sharon) NN
Impatiens capensis (Orange Balsom) NN
Iris spuria (Blue Iris) NN
Lilium martagon (Martagon Lily) NN
Lilium pyrenaicum (Pyrenean Lily) NN
Linaria purpurea (Purple Toadflax) NN
Lunaria annua (Honesty) NN
Lupinus arboreus (Tree Lupin) NN
Lysichiton americanus (American Skunk Cabbage) NN
Medicago sativa (Lucerne) NN
Melilotus alba (White Melilot) NN
Oxalis corniculata (Procumbent Yellow Sorrel) NN
Paeonia mascula (Peony) NN
Phacelia tanacetifolia (Phacelia) NN
Solidago canadensis (Canadian Goldenrod) NN
Symphyotrichum species and hybrids (Michaelmas Daisies) NN
Stachys alpina (Limestone Woundwort) NN
Tulipa saxatalis (Cretan Tulip) NN
Vinca major (Greater Periwinkle) NN

NN = Naturalised Neophyte

Native Weed Species

The subjectivity of an assessment of weediness may sometimes unfairly tarnish a perfectly harmless and beneficial native plant species with a bad name. Take, for example, our hero plant, Dandelion (*Taraxacum officinale*), a common native species of meadows, pasture, amenity grassland and lawns, considered a weed by many farmers, green keepers and gardeners. Its only sin is that it occupies space that could be growing grass. But as well as telling the time and looking pretty, Dandelions contribute to the environment and ecosystems in many good ways:

- The seeds, leaves, flowers and roots are palatable and nourishing.
- It is deep rooting and drought tolerant, drawing water and nutrients from low down.
- It is one of the top five native species for providing pollinator services.
- It removes CO_2 from the atmosphere and releases refreshing O_2.
- It is a detoxifying herb.
- An essence may be extracted from the roots and taken as an alternative to coffee.
- It is a hero.

Our native Dandelion (*Taraxacum officinale*).

The Weeds Act 1959

UK Public General Acts: 1959 c54

The Weeds Act 1959 lists five native species that are considered to be injurious to farming and horticulture. The Act requires landowners and occupiers to prevent these species from spreading to neighbouring land. It is not an offence to have these species growing on your land, but it is an offence if they are shedding seed too close to your neighbours' property. The seed of Dock disperses a few metres at most, that of Thistle perhaps 10m (30ft), and Ragwort rather more. A cordon of 25m (80ft) is usually sufficient to prevent spread to neighbouring land.

WEEDS ACT SPECIES

Cirsium arvense (Creeping Thistle) N
Cirsium vulgare (Spear Thistle) N
Rumex crispus (Curled Dock) N
Rumex obtusifolius (Broad-Leaved Dock) N
Senecio jacobea (Ragwort) N

N = Native

Our native Ragwort (*Jacobaea vulgaris*) can be a serious pest of agricultural grasslands.

Other Native Weeds

In the British Isles, Bracken (*Pteridium aquilinum*) is probably the most abundant, most harmful and most invasive of our native species, even more so than most invasive alien species. It is ubiquitous across the British Isles, and cosmopolitan world-wide, producing dense and expanding deciduous stands of tall fronds that easily out-compete and replace species-rich plant communities, harming biodiversity and ecosystem function countrywide, as well as being a serious agricultural and horticultural weed (Marrs and Watt, 2006; Milligan et al., 2015). A common feature of highly invasive species is that they may be toxic and/or unpalatable to livestock, and this is the case with Bracken, giving it a selective advantage in grazed landscapes. Bracken foliage and roots are also toxic to us, and the spores released in midsummer may be carcinogenic.

Most native species that have, by tradition, been thought of as weeds actually do little harm. However, a moderate number are injurious and/or responsible for economic harm to agriculture, horticulture and forestry.

NATIVE AND ARCHAEOPHYTE WEEDS

Examples of some native and archaeophyte weed species that may be injurious and/or may do economic harm, or have the potential to do so.

Aegopodium podagraria (Ground Elder) A
A difficult-to-control rhizomatous perennial weed of horticulture and gardens.

Alopecurus myosuroides (Black Grass) A
A winter annual grass weed of arable cropping that is difficult to control.

Arctium lappa (Burdock) A
A majestic plant, but burrs tangle in fleeces and barbed pappus bristles can enter the eyes.

Arrhenatherum elatius (False Oat Grass) N
An invasive smothering grass that becomes dominant on unmanaged grasslands.

Atropa bella-donna (Deadly Nightshade) N
A very poisonous species.

Chenopodium album (Fat Hen) N
A competitive spring annual weed of rich horticultural soils and gardens.

Conium maculatum (Hemlock) A
A toxic wayside biennial weed.

Convolvulus arvensis (Field Bindweed) N
A difficult-to-control rhizomatous perennial of grassland, horticulture and gardens.

Equisetum arvense (Field Horse-Tail) N
Another difficult-to-control rhizomatous perennial of grassland, horticulture and gardens.

Lolium perenne (Perennial Ryegrass) N
Invasive and competitive on richer soils.

Oenanthe crocata (Hemlock Water-Dropwort) N
A highly toxic wetland and waterside perennial.

Persicaria maculosa (Redshank) N
A competitive smothering weed of cultivated ground.

Polygonum aviculare (Knotgrass) N
A competitive smothering weed of cultivated ground.

Pteridium aquilinum (Bracken) N
An invasive and toxic perennial fern.

Ranunculus repens (Creeping Buttercup) N
A smothering creeping perennial of richer and damp soils.

Shedonorus pratensis (Meadow Fescue) N
A robust competitive grassland perennial.

Sinapis arvensis (Charlock) A
A phytotoxic spring annual weed with a long-lived soil seed bank.

Urtica dioica (Stinging Nettle) N
A robust invasive perennial, sometimes dominant on richer soils.

Vulpia myuros (Rat's-Tail Fescue) A
An emerging herbicide resistant winter annual of arable cropping and disturbed ground.

N = Native, A = Archaeophyte

Bracken (*Pteridium aquilinum*) can be harmful to both agricultural production and biodiversity.

CHAPTER 5

NATURAL CAPITAL AND ECOSYSTEM SERVICES

The concept of natural capital and ecosystem services provides a structure for calculating the monetary value of a unit quantity of natural biodiversity. Natural capital is the fixed asset value of a chunk of nature. For example, it may be a meadow, a wood or a pond, and it may be pre-existing, restored or created. The total value of UK natural capital in 2021 is estimated to be £1.5 trillion, even at this low point for nature. Ecosystem services are the value of the annual output of goods and services from the natural capital, and the annual output value of ecosystem services for

North Meadow National Nature Reserve.

the UK, estimated for 2021, is £47 billion (ONS, 2023). Looking at nature in this way allows a value to be put on nature (Farrell et al., 2022). The base value is calculated on evidence and data, and the current market value is determined by open trading of biodiversity and environmental credits.

North Meadow National Nature Reserve (NNR), a short walk from the market town of Cricklade in Wiltshire, is a good example. It is probably the best surviving lowland meadow in Britain, and extends to around 44.5ha (110 acres). It is a Lammas Floodplain MG4 meadow. Calculations of the monetary value of the ecosystem services provided by North Meadow give an estimated benefit for the carbon sequestered below ground of £32,324, an estimated annual benefit of £88,971 for other environmental ecosystem services, and an annual benefit of £5,814 for agricultural output (Lawson et al., 2018). The very high value of environmental ecosystem services is a feature of all species-rich native meadows and grasslands, whether existing, restored or created.

Table 1: The monetary value per year for different ecosystem services provided by North Meadow, Cricklade (reproduced from Lawson et al., 2018).

Benefit provided by floodplain	Description of the service delivering the benefit	North Meadow quantities	Value per unit	North Meadow total value
Food	Agriculture; crop and livestock production Hay values	Hay yield 4 t ha^{-1} yr^{-1}	£75 gross, £35 net	£5,439
	Grazing land value	0.4 LU ha^{-1}	£2.50 LU^{-1} week^{-1}	£375
Climate Regulation	Carbon sequestration (t/c/ha/yr)	Literature is inconclusive	Not conclusive	£0
Climate Regulation	Carbon storage below ground (soil, t c ha^{-1})	Soil carbon = 3 t ha^{-1} = 11 tCO$_2$e ha^{-1}	£66 tCO$_2$e^{-1} (DECC non-traded carbon price)	£32,324 (asset)
	Carbon storage (above ground tC ha^{-1})	Variable with season, no long-term store		£0
Pollination	Habitat for pollinating insects	44.4ha	£29.45 ha^{-1}	£1,308
Water quality	Sediment trapping	0.8 m^3 ha^{-1}	£13.83 m^{-3}	£491
Air quality	Removal of atmospheric pollutants	No data		
Natural Hazard Regulation	Flood storage (above ground)	44.4ha	£197	£8,746
Biodiversity	Species-rich habitats – high diversity and rare species	44.4ha	£499 ha^{-1}	£22,156
Cultural history	Strong 'sense of place' and social history	44.4ha of historic landscape	£203.4 ha^{-1}	£9,013
Aesthetic	Enhancement of the landscape, intrinsic appeal	No data		
Recreation	Enjoyment of the outdoors	15,000 visitors yr^{-1}	X £500 ha^{-1} yr^{-1}	£22,200
Health		2km of path with 50m wide buffer either side = 20 ha	£433 ha^{-1}	£19,225

Ecosystem Services

The major ecosystem services provided by existing, restored and created meadows may be summarised as set out in the sections below.

Habitat Provision

Green plants are the primary producers in a meadow ecosystem. They in turn support a wide range of other organisms, from the simplest to the most complex. These will include viruses, bacteria, algae, fungi, lichens, mosses and liverworts, ferns, invertebrates, amphibians, reptiles, birds and mammals (including humans). Meadows provide a habitat and the resources to support all these life forms.

Conservation of Biodiversity

Meadows are biodiversity rich. They are a living genetic resource for now and into the future. We cannot tell today what may be of utility tomorrow. Only 6.2 per cent of higher plant species worldwide have so far contributed to products of use to our species (Kew, SOTWP, 2020). We rely on plants directly for food, medicines, fuel, materials and more. Our future needs are unknown. The Convention on Biological Diversity, Rio de Janeiro (1993) was signed by 190 countries pledging 'the preservation of biological diversity, the sustainable use of its components, and the fair and equitable sharing of the benefits from the use of genetic resources'.

Facilitating Ecological Processes

Meadows are a dynamic living ecosystem of biodiversity, with species, individuals and the environment together interacting across space and time. This engine of interacting processes generates healthy genetic change within ecosystems and delivers the ecosystem services we so value.

Carbon Dioxide Removal

Earth's atmospheric CO_2 concentration was 3,000–9,000ppm 500 million years ago, dropping to below 300ppm at the beginning of the Pleistocene (1 million years ago) and staying at that equitable concentration until 1950 (it is now rising to 400ppm and more). Photosynthetic plants have been removing CO_2 from the atmosphere, and releasing refreshing O_2, for at least 2 billion years. All the plant origin sequestered carbon across the world is ultimately derived from photosynthesis, fixed below ground in hydrocarbon deposits and biomass, and in a good proportion of all calcium carbonate deposits (chalk and limestone). Meadows photosynthesise year round, capturing CO_2 from the atmosphere and locking it into organic compounds. Most carbon sequestration in meadows is below ground, visible as an accumulation of humus. Net carbon gain is greatest during the transition period when meadows are created on former arable land, and during the restoration phase when shallow-rooted pure grass swards are restored to deep-rooting, species-rich meadows (Lange et al., 2020; Cong et al., 2014). The accumulation of this below-ground biomass of fixed carbon is a very important ecosystem service.

When civilisation releases more CO_2 to the atmosphere than vegetation is able to remove by photosynthesis, the concentration of atmospheric CO_2 rises, as do global temperatures. This in turn increases the frequency of wild fires, reducing vegetation cover further and releasing more CO_2. This, in addition to desertification (caused by overgrazing and cultivation), forest removal for cropping and grazing, soil erosion, and removal of vegetation for development, all combine to reduce the global stock of wild vegetation working to remove atmospheric CO_2 and replenish vital O_2. In this way, a green planet turns brown and increasingly burns. Reducing CO_2 emissions and restoring the global photosynthetic capacity of nature, together, will slow global warming and eventually reverse it. A good reason to sow a meadow.

Pollinator Services

Across the British Isles pollinator populations have declined sharply since the 1950s (Steele et al., 2019). In a meadow, the invertebrate species that facilitate the pollination of flower species (the adults of butterflies, moths, bees, flies) are a hugely important constituent of the meadow ecosystem. Grasses are mostly wind pollinated: they produce powder-like pollen and have no need to form attractive flowers or produce nectar. In contrast, the 'wild flowers' (forbs) of a meadow are mostly insect pollinated. Together, flower species and pollinating insect species have co-evolved complex, mutually beneficial relationships with each other. Some of these relationships are very species specific, while others are more general. Normally, the

ABOVE: Meadow Brown butterfly (*Maniola jurtina*) on the rayed form of Common Knapweed (*Centaurea nigra*).

LEFT: Peacock butterfly (*Aglais io*) on Wild Marjoram (*Origanum vulgare*).

flowers attract pollinators using visual and/or scent cues, and reward the pollinator with nectar and/or sticky edible pollen. In consequence, both pollinator and wild flower populations thrive by mutual benefit (Nichols et al., 2019). Species-rich meadows are exceptional for providing pollinator services. When these services are measured as the quantity of insect numbers supported and the quantity of nectar and sticky pollen produced, species-rich meadows are far superior compared to gardens, parks, woodland and farmland (Hicks et al., 2016).

Pollinators are essential for the production of many agricultural and horticultural crops worldwide (Smith et al., 2022), the value of which is estimated at £200bn–£450bn annually (JNCC, 2022). Fruit set in crops such as strawberry, gooseberry, apple and pear all require insect pollination. It is a mutualistic relationship. The crop, in return for fruit set, nourishes the pollinator with

Green-veined White butterfly (*Pieris napi*) on Meadow Buttercup (*Ranunculus acris*).

Comma butterfly (*Polygonia c-album*) on Nettle (*Urtica dioica*).

Meadow Brown butterfly (*Maniola jurtina*) and Five-spot Burnet moth (*Zygaena trifolii*) on Field Scabious (*Knautia arvensis*).

energy-rich nectar and protein-rich pollen. It is important to foster this mutualism, both in the agricultural setting and in the wild. Bringing nature and cropping together can provide continuity of nourishment for pollinators throughout the flight season. However, large areas of single crops may tempt pollinators away from native plant communities when the crop is flowering, reducing seed set in the wild. A balance is required for mutualism to function sustainably.

Clean Air

Meadows clean the air passing over by filtering out particulates, dropping these to the ground, where microbes will metabolise harmful constituents and deposit the residue in the soil profile. In the UK, the natural environment is estimated to have removed 869,166 tons of air pollutants in 2021, at an annual value of around £2.5 billion to the nation (ONS, 2023).

Also important, photosynthesising plants in a meadow remove harmful CO_2 from the air passing over, fixing the captured carbon in biomass, and releasing refreshing oxygen back to the atmosphere. The regulation of both O_2 and CO_2 at equitable concentrations in the atmosphere are vital ecosystem services of wild vegetation.

Clean Water

Rain falling on a meadow falls through the vegetation into the soil profile, descending until it meets the water table, then flows sideways until it exits, cleaner, into the destination watercourse and catchment. On conventional agricultural land the water entering the catchment will carry a burden of leached pesticides, fertiliser, livestock waste and sediment. The value of 'clean water' ecosystem services can be calculated as the cost falling on water companies to remove toxic components from drinking water derived from polluted farmland sources.

Water Storage and Moderating the Storm Surge

Meadows moderate and smooth the extremes of the water cycle. Water falling as rain percolates slowly down through the soil profile, assisted by the intricate network of fine channels left by worms and roots. The above-ground vegetation of a meadow will significantly slow the downhill flow of surface water that can quickly accumulate during heavy rainstorms. On level floodplains beside watercourses meadows can accommodate and store huge quantities of water when a river bursts its banks. This filtering, slowing and storage of water is hugely important in reducing the 'storm surge' downstream, lessening the threat of flooding to settlements, property and life. The cost/benefit yield of water storage on floodplain meadows is around 1:9 (JNCC, 2022; Everard, 2018).

Soil Formation

Meadow ecosystems build soil, slowly over millennia. The older the meadow, the better the soil that lies beneath it. Over time, soil humus accumulates, essential nutrients accumulate locked up in biomass, crumb structure improves, and a complex stratified profile evolves. But the meadow can be destroyed by ploughing, releasing all this goodness to produce a few bumper crops of cereals, a fact not lost on mid-twentieth-century farmers.

Access and Recreation

Where public rights of way exist, or permissive rights are given, the health, well-being and educational benefits of access can be huge. This is another ecosystem service provided by meadows and wild surroundings.

Heritage Enhancement and Protection

Existing, restored and created meadows contribute to landscape quality by maintaining historic field patterns, traditional land use within these fields, and the structure of farm holdings. Also, meadows are an excellent land use for preserving below-ground archaeology, unharmed.

Agricultural Production

Agricultural output from land, principally the production of food, is an ecosystem service, no more and no less important than the delivery of environmental services. In our example of North Meadow NNR, the value of agricultural production is dwarfed by the huge value of the environmental ecosystem services provided. It is an exceptional site. Across the British Isles, calculations of natural capital and ecosystem services will tell us which land use is best. On high-quality agricultural land, the value of food production (when tempered by measures to reduce pollution) may outweigh the value of environmental ecosystem services derived from creating a meadow. On marginal land both food production and nature may benefit best by existing side by side in harmony. In the uplands and on poor ground, it may be of more net benefit to forego or reduce intensive agriculture than to continue to farm difficult land.

Sustainability

Sustainable land use is use that results in no net loss or harm to the constituents of the land, to nature and to the environment, and that can continue indefinitely without net harm. For too long we have exploited and harmed nature and the environment, with impunity, for profit and for convenience. We have cut down forests and hedges, dug peat, ploughed old grasslands, drained wetlands, burnt fossil fuels, mined minerals, chemical farmed, and polluted air, water and soil. The cost of these harmful exploitative activities can be calculated by measuring the lost value of ecosystem services when nature is harmed, the cost to society for putting right the environmental damage caused, and the cost of government subsidies supporting these exploitative activities. Combining these three costs puts a value to the debt that should fall on the polluter.

The government in its '25 Year Plan' (UK Government 2018, 2023) supported 'the polluter pays' principle. If applied in full as a pollution added tax (PAT), organic food becomes cheaper than chemical farmed food, and green energy becomes cheaper than fossil fuel, overnight, and the pollution tax revenue could fund restoration and repair of the environment.

CHAPTER 6

OPPORTUNITIES FOR ECOSYSTEM RESTORATION AND CREATION

In this chapter we look at different land uses and the opportunities they present for nature recovery. For convenience we divide land into four general categories: (1) the wild, (2) farmland, (3) gardens and parks, and (4) city, road and rail. Of these, the least polluted and the most biodiverse are the best wild areas. The most polluted and the least biodiverse is intensively managed farmland. Gardens, parks, city, road and rail fall in between, often being less polluted and more biodiverse than intensive farmland, but not as nature rich as the wild.

Restored lowland neutral meadow.

LAND USE IN ENGLAND, 2022 (GOV.UK, 2022)

- 63.1 per cent agriculture
- 20.1 per cent forestry, open land and water
- 8.7 per cent developed land including residential
- 4.9 per cent residential gardens
- 3.2 per cent miscellaneous uses

The Wild

'The wild' are those places in nature where the hand of nature, not the hand of woman or man, has dominant control over the composition of the vegetation and the functioning of ecosystems. By this definition, a species-rich unimproved (agriculturally) meadow is wild, even if cut by the farmer for hay and the aftermath grazed by livestock.

The best examples of surviving wild vegetation are thinly spread in isolated pockets across the British Isles. Statutory protection, for reasons of biodiversity or geology, is given to only 6.1 per cent of the land in England. Most of such land is designated as Sites of Special Scientific Interest (SSSI), and may also be designated Special Areas of Conservation (SAC), a National Nature Reserve (NNR), a Super National Nature Reserve (SNNR) or a RAMSAR site, the latter having importance for the conservation of wetland biodiversity. These sites have legal protection.

The Environmental Impact Assessment (Agriculture) Regulations (2014) add another layer of protective legislation, and these measures seek to protect all wild and semi-natural vegetation by requiring prior approval for agricultural intensification by chemical and/or physical means. These regulations only apply to fields or areas greater than 2ha (5 acres), smaller areas not having any protection. This cut-off is unfortunate, as it removes protection from smaller pockets of biodiversity, which, following widespread agricultural intensification, is all we have left over much of the landscape.

Non-statutory recognition is given to Local Wildlife Sites and Sites of Nature Conservation Interest (SNCI). These sites account for 5.2 per cent of land in England. Ancient woodlands, also with no statutory protection (unless SSSIs), account for 2.7 per cent of England. Public bodies have a duty of regard for these sites when carrying out their functions.

At a landscape scale, National Parks (9.1 per cent of England) and Areas of Outstanding Natural Beauty (14.4 per cent of England) are protected primarily for scenic beauty, not for biodiversity, other than may be incidental to scenic beauty.

Garibaldi *et al* (2020) calculate that working landscapes (farmland and forests) require at least 20 per cent of native wild habitat to allow these landscapes to function in a productive, beneficial and healthy way. The working landscape of the British Isles is currently well short of this minimum area of native habitat, having lost 97 per cent of species-rich meadows. Professor Lawton, in *Making Space for Nature* (Lawton et al., 2010), proposed resilient ecological networks, making wild sites larger, more robust, more resilient and better protected, and also better connected by restoring and creating wild corridors and stepping-stones at a landscape scale.

Making Wild Sites Larger to Improve the Resilience of Meadow Ecosystems

Some remnant wild meadows are partly or completely enclosed by housing estates and commercial development, and offer little opportunity for expansion. An example is Hucclecote Meadows SSSI, near Gloucester. However, most unimproved meadows (agriculturally unimproved) are surrounded by farmland or open countryside, and the nature of land tenure nearby may, in some cases, provide opportunities for expanding the existing wild meadow. Brush-harvesting methodology is the best way of going about this, using the original meadow as the seed donor site. This approach ensures that the scientific and heritage values of the original site are preserved and are not challenged by plants introduced from elsewhere.

There are many examples where this approach has been used successfully. At North Meadow NNR, a floodplain meadow near Cricklade, seed that was brush-harvested from the main meadow was sown (in 2012) on an adjacent former arable field, and floodplain meadow vegetation is now gradually developing, with Great Burnet (*Sanguisorba officinalis*), Meadowsweet (*Filipendula ulmaria*), Snake's-head Fritillary (*Fritillaria meleagris*) and Pepper Saxifrage (*Silaum silaus*) all becoming established.

Size is important for resilience. The larger the meadow, and the larger the populations of individual species in that meadow, the less are the risks from inbreeding depression. In out-crossing species, small isolated populations of less than around fifty to sixty individuals may experience inbreeding depression, with the population falling inevitably into the *extinction vortex,* from which there is no escape unaided. Yearly revisiting small populations of a rare species in a favourite meadow is, over time, often increasingly poorly rewarded, until none of the original population is found. In this way, small isolated populations of rare species are lost.

Creating Corridors and Stepping-stones to Facilitate Gene Flow between Wild Populations

Connectivity is important (Auffret et al., 2017). Small isolated meadow communities may lose genetic variation over time from inbreeding and natural selection. As genetic diversity declines, so does the ability of the community to respond to changing climate, weather, management, and pests and diseases. A meadow that is not genetically diverse and robust is less able to respond to changes and challenges.

Genetic robustness of small meadow communities can be maintained over time if the loss of genetic variation by inbreeding and natural selection is replaced by genotype and gene flow into the community. A good example from the animal kingdom is the Florida Panther (Johnson et al., 2010). A remnant, isolated and very small population of this species, consisting of twenty-six individuals, displayed an alarming suite of inbreeding depression symptoms, including tail kink. The population was predicted to become extinct in two decades if it remained isolated. Eight wild-caught female Texas Panthers (a related species) were translocated and introduced in 1995, and as a result of cross-breeding, genetic diversity in the Florida Panther population increased, the symptoms of inbreeding depression declined, and the population size increased threefold. This is a textbook example of genetic rescue.

In the natural plant world, genes move from place to place in seeds (as a complete intact genotype), in pollen (as a random selection of genes), and in spores in lower plants. The rate of gene flow (quantity of genes × distance/generation time) varies enormously between species depending on life cycle, pollination mechanism and seed-dispersal traits. Stepping-stones and corridors can be designed around these features to ensure maximum desired gene flow (Geert et al., 2010).

A plant shedding seed and pollen will cast an exponential seed and pollen shadow, with the greatest quantities usually going only a very short distance from the mother plant. The shadow also has a tail, which for some species may be very long. In sexual species, genotype and gene flow over distance alternates repeatedly and stepwise between pollen and seed, as of necessity.

Average seed and pollen dispersal distances can give an indication of gene-flow rates along linear friendly habitat. Inevitably, rare, very long-distance seed-dispersal events occur, and can greatly extend the range of a species in one leap. This is fine for selfing and apomictic species. But lone founder individuals of outbreeding species require equally long-distance pollen transfer to set seed. This would be a rare event for an outbreeding species, and long-distance founder individuals usually see out their time without reproducing.

It is tempting to visualise life-saving novel genes hopping from plant to plant along habitat corridors and stepping-stones. In fact, each pollination event carrying a novel gene must be followed by a seed cycle (which, for meadow species, will be one to eight years) before the novel gene can

LIFE-CYCLE FEATURES AND GENE-FLOW RATES

Apomictic species: No functional pollen dispersal therefore no gene flow by pollen. Genotype flow by seed dispersal, but genes locked in. *Taraxacum officinale, Hieracium sp.*

Selfing species: No functional pollen dispersal therefore no gene flow by pollen. Genotype flow by seed dispersal, but genes locked in.

Outbreeding species with no seed-dispersal adaptations: Moderate gene flow by pollen, ± 50m (164ft). Low genotype flow by seed, ± 20cm (8in). *Primula veris, Centaurea nigra.*

Outbreeding species with short distance seed-dispersal features: Moderate gene flow by pollen, ± 50m (164ft). Moderate genotype flow by seed, ± 2m (7ft). *Geranium pratense, Lotus corniculatus.*

Outbreeding species with long-distance seed-dispersal adaptations: Moderate gene flow by pollen, ± 50m (164ft). High genotype flow by seed, ± 100m (330ft). *Leontodon hispidus, Agrimonia eupatoria.*

Outbreeding species with dust seed: Moderate gene flow by pollen, ± 50m (164ft). Very high genotype flow by seed, ± 1km (0.6 mile). Orchid species.

move to the next individual by pollen transfer. Most meadow species are outbreeding, have a generation cycle of around five years, have limited seed dispersal, and an average pollen transfer distance of ± 50m (164ft). This gives an unaided gene-flow rate in congenial habitat of around 10m (33ft)/year.

Before modern-day farming, meadows were cut for hay, and some of the hay was traded between farms, creating long-distance seed dispersal and gene flow. It was also common practice to collect feeder and barn sweepings to patch up poached gateways, and barn hay was often winter-thrashed by seed merchants for hayseeds for selling on for sowing down new meadows on former arable land. These days most forage is ensiled by wrapping large round and square bales. This kills the seeds within when the bale heats and self-sterilises. The creation of habitat corridors and stepping-stones may be taken as an opportunity to use seed propagation material from diverse genetic sources, and so assist the long-distance gene flow in meadow species, formally assisted by the trade in hay and hayseeds. The corridors need not be wide, or the stepping-stones large. A hedge with a 2m (7ft) wild margin each side, and a further 5–10m (16–33ft) each side chemical free, stretching across the landscape, would be sufficient.

Genetic Rescue

But how to introduce novel genes into a genetically impoverished, but important remnant meadow without introducing unwanted species and ecotypes? This may be achieved by establishing a small 'genetic donor meadow' close to the site of the genetically impoverished meadow. The distance between the two must be greater than the seed dispersal distance of the species of concern, but less than the pollen transfer distance of those species. The effective separation distance will vary from species to species depending on life cycle, pollen transfer distance and seed dispersal distance. For most species, the effective separation distance will be around 50m (164ft). In this way genes (via pollen) can enter the sensitive site, but not genotypes (as seed). Entering genes will increase the genetic diversity of the target meadow species, lower inbreeding coefficients, and introduce beneficial genes that will become incorporated, by natural selection and introgression, into the genetic make-up of the genetically impoverished community.

For large healthy meadows of high conservation value it is important to continue beneficial management and to prevent, as far as possible, the arrival and establishment of non-native species and non-local ecotypes. These important sites are valuable in providing a high-quality brush-harvested seed source for meadow creation and restoration elsewhere.

Restored calcareous meadow ten years after scarification and overseeding of improved agricultural grassland.

SUMMARY ECOSYSTEM RESTORATION AND CREATION IN THE WILD

- Increase the genetic diversity, population size and geographic size of isolated remnant ecosystems using seed from adjacent donor vegetation, or a nearby wild local source.
- Increase gene flow at landscape scale by creating and restoring habitat corridors and stepping-stones linking communities.
- Allow vegetation to revert to nature naturally by removing grazing livestock from the uplands and ceasing to cultivate marginal arable land.

Farmland

Farming is the major land use in Britain and Ireland, accounting for around 70 per cent of all land. Not surprisingly, agriculture has had a dominant influence on nature and the landscape, past and distant past. Farming was at first benign, co-existing happily and productively with and alongside nature. It is now industrial and chemical. The change from wild lands to farmland was gradual, but gathered pace as methods became more intensive and as the population of the British Isles grew exponentially 1800–2000. The two World Wars and food rationing exposed our food insecurity, a lesson hard learnt. In response, the Agricultural Act 1947 put in place policies to modernise farming, increase home production of food, and reduce reliance on imports. Our farmers rose to the challenge by greatly increasing the production of home-grown food. Of course, this could not be done without grubbing out hedges and ploughing up meadows, and looking back we can understand the necessity.

More recently, the Common Agricultural Policy (CAP) has supported farming within the European Union, and continues to do so. Initially, commodity subsidies were put in place to support the production of high-calorie foods (such as sugar, flour, butter), and this led to over-production of these foods, and under-production of unsubsidised salad, fruit and vegetables. To curb over-production, commodity subsidies were replaced by area payments, a generous form of income support for farmers, especially those with large estates. Of interest to us, the EU, recognising the harm done to nature by modern farming, introduced under CAP, options for agri-environment schemes, as a devolved matter. Although the adoption of these schemes across Europe has been lukewarm, it was a start.

AGRICULTURAL ACT 1947

- Established state plant-breeding institutions to produce high-yielding cultivars.
- Established an advisory service (ADAS), agricultural colleges and experimental farms.
- Set price guarantees for eleven major agricultural products.
- Issued grants to drain and plough up meadows and seed with agricultural grasses.
- Issued grants to grub out hedges and make fields larger.
- Implemented measures to amalgamate small farms into larger units to gain economies of scale.
- These measures together achieved the modernisation of farming.

COMMON AGRICULTURAL POLICY

- Production subsidised by headage payments, intervention, quotas and tariffs.
- These measures resulted in unsold butter mountains and wine lakes in Europe, while people in poorer regions of the world were starving.
- Later, area-based income support replaced most commodity subsidies in a bid to curb over-production.
- These measures together achieved low market prices for food and plentiful supply.

The UK has now come out of the CAP, but for largely historic and political reasons, additional domestic policies remain in place, giving favoured status to the UK agricultural sector. This is of interest to us because intensive farming is the opposing land-use option to nature recovery, and landowners will understandably make land-use decisions based on prevailing policy and revenue opportunities.

John Bowers in his study of post-war agricultural policy observed: 'The lasting achievement of post-war (agricultural) policy has been the changes it wrought in the landscape and the natural environment. By common consent these have been deleterious, if not disastrous' (Bowers, 1985).

CURRENT MEASURES BENEFITTING THE UK FARMING SECTOR

- Exemption from full planning control, business rates, inheritance tax on most agricultural property, and zero VAT on most farm outputs.
- Exemption from the control of diffuse pollution from herbicides, pesticides, fertiliser, veterinary products, livestock emissions, slurry and sediment.

Chemical Farming

Chemical farming is bad news for nature. All of nature's life forms are harmed in one way or another by the agricultural use, application and release (to the environment) of harmful synthetic chemical components of sewage sludge, fertiliser, insecticides, herbicides, nematocides, fungicides, rodenticides, molluscicides, antibiotics, and so on. These harmful novel Frankenstein chemicals are the product of the chemist's laboratory, and are not naturally occurring. Their lethal potency derives from their biological novelty. Life forms have evolved, over millions of years, a tolerance of and a resistance to naturally occurring harmful environmental chemicals such as salt and heavy metals. They have no inbuilt tolerance of, or resistance to, the novel poisons produced in the laboratory.

For example, synthetic neonicotinoids are one of the most harmful. These are highly potent novel insecticides that leak from the crops to which they are applied into the adjacent environment, where they can persist for months or longer. Long banned in the EU, their use has been permitted in the UK on Sugar Beet (*Beta vulgaris*) and Oilseed Rape (*Brassica napus*) in the three years following Brexit. One teaspoon full of the neonicotinoid insecticide thiamethoxam is sufficient to kill 1.25 billion bees. Crops of Oilseed Rape produce abundant nectar and pollen, and they require pollinating bees for good seed set and top yields – but a neonicotinoid-treated crop simultaneously delivers a debilitating and often lethal dose of insecticide to the pollinator in the nectar and pollen it consumes. Also, 'wild flower' crop margins, while appearing to blend nature with farming, can lure pollinators away from safe habitat to the leaky edge of neonicotinoid-treated crops.

Chemical farming has also harmed the arable flora and the grassland flora of the British Isles. The modern arable crop is completely weed free, and the application of herbicides to grassland to control Weeds Act species has almost completely removed wild flowers from pasture and mowing ground on most farms.

Of course, not all farms are intensive chemical farms. Some farmers prefer not to use poisons. Organic farming is the greenest alternative, producing food mostly without the use, application and release of harmful synthetic chemicals.

Plant Breeding

Chemical farming has long passed the sustainability threshold, and increased chemical sophistication and high-tech precision application will never make chemical farming fully sustainable. Plant breeding could help here.

State plant-breeding institutions have, since the last war, bred and released numerous very high-yielding agricultural and horticultural cultivars. These have been enthusiastically adopted by farmers and growers, and are the main factor responsible for the large post-war increases in crop yields. The full potential of these cultivars is dependent on modern cropping cultural programmes requiring short rotations and high inputs of pesticides, herbicides and artificial fertilisers. Of course, these cultural programmes have had an unfortunate and harmful environmental impact, but they have produced the food we need.

In the UK, around 29 per cent of land is improved grassland sown to high-yielding forage cultivars of, mostly, Perennial Ryegrass (*Lolium perenne*) and Italian Ryegrass (*Lolium multiflorum*). The arable land is sown to Wheat (*Triticum aestivum*), Barley (*Hordeum vulgare*), Oil-seed Rape (*Brassica napus*) and others, and accounts for around 23 per cent of land cover. So currently, around half of the British Isles is sown to monocultures of crops, all requiring inputs of environmentally harmful chemicals if grown by conventional means for maximum yield.

Before they can be marketed, new agricultural and horticultural cultivars must be registered for plant variety rights (PVR). This registration requires that any new cultivar is 'uniform, stable and distinct'. The requirement is of necessity, otherwise it would not be possible to describe a cultivar in a way that distinguished it from all others, and in a way that demonstrated that the breeder had created something new and unique. Most agricultural cultivars are seed propagated, for example wheat, barley, oats, linseed, field beans, peas, oil-seed rape, sugar beat and perennial ryegrass.

Therefore to be 'uniform and stable', seed-propagated crops must be without genetic variation between and within seeds. In other words, they must be completely true breeding, completely homozygous, and hence, completely without genetic diversity. In consequence, the cultivars on our farmland, unlike wild plants, are completely uniform. They are unable to benefit from the vigour of heterosis, and are unable to respond (by natural selection) to environmental change in space and time, unlike wild plants. Therefore in the face of climate change or new disease outbreaks, the cultivar may succumb and the plant breeder will be required to return to the breeders' field to produce new, updated cultivars in response to this change. Of course, this is a never-ending process until we can find a better way.

Inbred lines have an inherent weakness. They tend to have low vigour, low resistance to pests and diseases, low competitive ability against weeds, and low yield unless fed fertiliser. Modern inbred grain crops are reliant on high inputs of synthetic chemicals to overcome the inherent features of inbreeding. Chemical compensation is necessary if grain crops are to yield well.

Having largely met the challenge of post-war food security, the challenge today is to maintain that food security while restoring the environment and ensuring that we, and the environment, have a secure future. Plant breeding may again open a path to a better future, as it has done in the past.

Traditional plant breeding consists of controlled crossing between selected parents, planting out the progeny and making further selections. This can continue for many generations. When a potential new cultivar is identified, pure breeding stock enters field multiplication until commercial scale quantities are produced. This all takes many years. Genomics-assisted breeding can speed up the process and can help incorporate genetic material from undreamt-of sources.

Yield, and resilience gain in Wheat (*Triticum aestivum*), Barley (*Hordeum vulgare*), Rice (*Oryza sativa*) and Maize (*Zea mays*) by conventional breeding methods, is increasingly difficult. These are open-pollinated, or self-pollinated crops, maintained as inbred seed lines, and they have an inherent weakness: they are inbred. Darwin (1876) observed that inbreeding created a lack of vigour and that outbreeding generated heterosis. The heterosis of the F1 generation (the first hybrid generation between two inbred lines) is normally manifest as a substantial jump in yield and vigour, compared to the parents. This is evident in the widely grown and popular F1 hybrid cultivars of Maize (generated by detasselling the female parent of the hybrid), and the widely grown F1 hybrid cultivars of Rice (generated by using a male sterile female parent) (Li et al., 2023). Wheat (producing around 20 per cent of global calorie intake) and Barley are still cultivated as inbred seed lines. Both crop species have proven difficult to emasculate (by incision or genetics), and are not sufficiently responsive to chemical castration. We seem a long way from having global-scale cultivation of F1 hybrid Wheat and Barley produced by conventional breeding (Paril et al., 2023).

Lone Common Poppy (*Papaver rhoeas*) surviving on the edge of a field of Barley (*Hordeum vulgare*).

Wheat (*Triticum aestivum*).

Wheat is a self-pollinating species and has a complex allopolyploid genome (2n=6x=42). Some heterosis (not much) is locked into polyploid repetition and linked gene clusters, but present-day cultivars are still pure breeding inbred lines. Hand-performed crosses between different cultivars of Wheat will produce F1 hybrid individuals that exhibit hybrid vigour, but that heterosis is lost on selfing at the first and subsequent generations. There seems no easy way to lock heterosis into Wheat cultivars. However, breeders are yet to fully explore the native wild plant biodiversity of Britain and Ireland. It is a huge natural resource of genetic diversity.

Take our hero plant, the Dandelion (*Taraxacum officinale*): it has a completely different lifestyle to that of Wheat. It is an apomictic species, one of around 400 wild apomictic species worldwide, and it has opted, we assume, to lock in heterosis following a historic F1 hybridisation event. Dandelion produces seed embryos by mitotic nuclear division, in place of the more normal meiotic reduction division and zygotic fusion, and, like some other apomictic species of the Asteraceae, it does not require pollination for endosperm development. All seeds produced by a Dandelion plant are genetically identical to each other and to the mother plant, and all also carry the apomixis genes and heterosis to the next generation unaltered.

Dandelion, although one of the top five native species for providing pollinator services (Hicks et al., 2016), does not require the services of pollinators, and will produce seed regardless. Apomixis is a life-cycle strategy that can lock in the hybrid vigour and the selective advantage of heterosis, and reproduce it unaltered through the seed line *ad infinitum*. Turning to Wheat, hand-performed crosses between different cultivars can produce F1 hybrid grains. Genomics may then be used to insert the genetics for apomixis into the F1 Wheat, having harvested the capability from a Dandelion, or another wild apomictic species, for example Goldilocks Buttercup, *Ranunculus auricomus* (Barke et al., 2018). Thus, hybrid vigour is locked in, the F1 Wheat genome is fixed, and it will reproduce faithfully down the seed line, passing on vigour, competitive ability, improved disease and pest resistance, and an increased ability to yield well, *ad infinitum* (Stokstad, 2023). Crucially for us, higher Wheat yields would free up more land for nature, and less use of agro-chemicals would reduce the harm done to the environment.

The next challenge, in a changing world, is to produce grain and forage crops that can adapt in the field to differing conditions between and within fields and locations, and to changes in weather and climate over time. The resilience of individual apomictic F1 hybrids is fixed, but a community composed of many different apomictic F1 hybrids would be able to respond to challenge and change. A blend of 100 F1 hybrids may be sufficient. The adaptive response (of the blend) in the field would be manifest as a change in the proportions of individual component F1 hybrids, but not in their genetics. Generations of farm-saved seed will, in short time, produce farm-specific well-adapted F1 apomictic blends of the crops grown.

Agri-environment Schemes

Agri-environment schemes come in many forms, and pay farmers and land owners for on-farm activities that benefit nature and the environment. Payments were initially based on 'income foregone', being the income lost for not farming the land intensively. Payments have edged up with increasing realisation that the value of the ecosystem services provided often exceeds the 'income foregone amount', sometimes substantially.

Agri-environment schemes are the main vehicle for delivery of ecosystem restoration and nature recovery in the farmed landscape. The first such scheme in England was the 'Environmentally Sensitive Area Scheme' (ESAs), which paid farmers for changes in land management, such

UPLAND HAY MEADOWS IN THE YORKSHIRE DALES

The species-rich upland hay meadows of the Yorkshire Dales, as elsewhere, are fragmented and degraded by agricultural intensification. Agri-environment schemes have attempted to restore these meadows by supporting the resumption of traditional management and low-input farming in the Dales. The results have been disappointing, with little evidence that management alone will restore these grasslands to species-rich meadows. Trials at Ingleborough NNR, monitored over ten to fifteen years, showed that plant diversity only significantly increased when the absent species were introduced as seed in combination with traditional management (Smith et al., 2002).

as reducing fertiliser application to upland hay meadows and cutting hay later in the summer. It was found that without introducing species as seed, management alone was not enough to significantly improve the biodiversity of semi-improved upland hay meadows (Smith et al., 2002).

Successor schemes to the ESA scheme have supported meadow restoration, combining beneficial management with seeding with native species, using brush-harvested seed, green hay, or special seed mixtures. These schemes have yielded biodiversity gain (Redhead et al., 2022; Pywell et al., 2012).

AGRI-ENVIRONMENT SCHEMES

- 1991–2004: Countryside Stewardship Scheme (CSS).
- 2005–2014: Environmental Stewardship, Higher Level and Entry Level Stewardship (HLS and ELS).
- 2015–present: Countryside Stewardship (CS), Mid Tier and Higher Tier Stewardship.

Land Use in the Farmed Landscape

Future land use in agricultural landscapes may be reasonably divided into three broad proposed categories: sparing, sharing and farming (Landis, 2017; Feniuk et al., 2019).

Sparing: Converting land to wildlife habitat and restored ecosystems. For example, in the uplands, reduced or zero grazing would allow the natural regeneration of wild ecosystems over extensive areas at present intensively sheep grazed.

Sharing: High nature value farming that produces both food and ecosystem services (Bullock et al., 2007; Isbell et al., 2017). For example, in the lowlands on marginal land it can be beneficial to combine food production with nature recovery (Brooker et al., *2021)*. Agri-environment schemes provide a framework for doing this.

Farming: Land farmed to produce high-yield, high-quality food with minimal harm to the environment should be encouraged where the conditions are best to do that: for example, in the lowlands on medium to fertile land. Here, farming should be supported and focused on high yields of high-quality food.

Habitat Creation and Restoration on the Farm

The opportunities for habitat creation and restoration on the farm are numerous. These will often be guided by options and prescriptions specified in agri-environment schemes. For example:

- Sow buffer wild-flower strips and margins (Westbury et al., 2018; Stroot et al., 2021).
- Restore species diversity to species-poor grassland and meadows (Pywell et al., 2007; Gamble et al., 2012).
- Create pollinator flower margins (Filipiak, 2019).
- Encourage (rare) annual weeds in arable cultivation (Storkey and Neve, 2018) by reduced crop density (Rotches-Ribalta et al., 2020), and zero or reduced chemical input cropping (Albrecht et al., 2016).
- Also wild up and make space for nature in orchards, wetlands, ponds and watercourses, rocks, scree and ravines, hedgerows, boundary walls, woodland and scrub, buildings, yards, tracks and hard standing, ruins, quarries, pits, muck heaps and dumps.

Conservation headland round the field edge of a conventional arable crop.

Gardens and Parks

Welcome nature in and make a difference: more biodiversity, cleaner air and water, healthy soil, more carbon capture, and less work. Every small patch turned wild makes a difference, gives nature a helping hand and a home. A wild windowbox will attract and nourish butterflies and bees. A wild pond, however small, soon attracts water boatmen, and a bush or hedge bearing autumn berries will feed hungry garden birds. As an entry point for the uncertain wild gardener, a small meadow of mostly non-native annuals is a good start. And then with confidence, it is but a small further step to go completely native and wild.

The conventional garden or park, although superficially green, is not as environmentally beneficial as it may otherwise be. The cultivation of the ground, the nurturing of alien species and cultivars, the weeding, the fertilising, the mowing, the spraying and the burning, all take their toll on nature and the environment. The following are a few of the easy measures, and adjustments, to conventional gardening and park management that may help save time, money and the environment:

Old-style cottage garden with a mix of native and alien species allowed to naturalise one with another.

- Reducing sprays and applications of insecticides, rodenticides, molluscicides, fungicides and herbicides is good for nature. These are all highly toxic in trace amounts (as they are intended to be), they hang around, and are quite mobile in the environment.
- Reducing the rate of application of artificial chemical fertilisers is good for the environment. Using garden compost is a sustainable alternative.
- Reduced frequency of mowing in the spring and summer can produce a transient flowering lawn attractive to pollinators.
- Green waste is better composted than burnt. Smoke from the garden bonfire is an unpleasant particulate pollutant, combustion releases CO_2 to the atmosphere, and the heat and flames are harmful to creatures. A compost heap, the bigger the better, can be a significant carbon store and may be used to offset unavoidable CO_2-releasing activities, such as essential air travel.
- Hand scything tall grass is better than strimming. It is quieter and not smelly, while sparing the glow-worm and the slow worm.
- Resist 'dead-heading' all the flowers. It removes the seed on which plant life cycles depend, and on which birds and other wildlife feed in winter. And saving home seed is the best way to maintain those favourite plants.

Non-native bedding plants on sale at a market in Soller, Majorca.

- Leave some dead vegetation standing over winter as a habitat and a hibernation spot for invertebrates and other creatures. It can be cleared away in early spring.
- Reduce weeding in the perennial border and vegetable patch, leaving the wild annuals (those harmless ones) to fill in the gaps.

All these good actions and inactions, in the garden and park, make it a better place for nature.

Ornamental Aliens

Ornamental cultivars of alien species are prized in the garden and glasshouse for their exotic looks, their large flowers, their bright colours, their diversity, their novelty, their exacting requirements, and because they look so different to our native wild flowers. It is not surprising that alien ornamentals are so popular with collectors of interesting specimens. We are a nation of proud and happy gardeners, justly rewarded by the pleasure of spending quality time immersed in caring for, and being surrounded by, the glorious beauty of our non-native floral creations.

However, conventional gardens and parks, for all their tidy beauty, are under-performing on natural capital and the provision of ecosystem services. Land occupied by alien ornamentals is not contributing to the restoration of native plant biodiversity. Small populations of non-reproducing alien ornamentals are not contributing to dynamic natural evolutionary processes. Space maintained between individuals and specimens in parks and gardens limits the between-plant interactions that are so important for ecosystem complexity and structure, this being the driving force of ecosystem service productivity. And maintaining cultivated populations of ornamental non-natives in gardens and parks risks invasive aliens spilling over into the wild.

Pollinators

The tender garden neophytes Lantana (*Lantana camara*) and Garden Lavender (*Lavendula angustifolia*), and the hardy naturalised garden neophyte Butterfly Bush (*Buddleja davidii*), are undeniably superb at attracting pollinators and are a welcome source of pollen and nectar. However, some alien ornamentals have pollen or nectar that is toxic to some of our native pollinators (Egan et al., 2016), and attracting native pollinators to the park or garden may disrupt complex mutualistic networks between native species (Arroyo-Correa et al., 2019), and may reduce the number and range of pollinators available for populations of native plants in the nearby wild. In addition, several studies have found that, when comparing like with like in our climate, native plants produce more pollen and more nectar than comparable alien ornamentals, and attract more pollinators than do alien species (Corbet et al., 2001; Salisbury et al., 2015). This is not surprising.

The relationship between plant and pollinator can be species specific, complex and mutualistic. This applies equally to native plant/native pollinator relationships and to ornamental alien plant/alien pollinator relationships, the latter species being brought to these islands leaving their co-evolved pollinator at home. When native pollinator meets alien ornamental it may be confusing for both.

Ornamental Escapees

Some alien ornamental species have jumped the garden fence and established reproducing and self-sustaining populations in the wild. Around 1,500 neophytes are naturalised

Rampion Bellflower (*Campanula rapunculoides*), a frequent non-native garden escapee found on wayside walls.

in the wild, many of which originated as garden escapees or garden throw-outs. A few of these are welcome additions to the wild flora of the British Isles, most are benign, and a few are seriously invasive, pushing out native species and over-running wild habitat, such as Rhododendron (*Rhododendron ponticum*) and Cotoneaster (*Cotoneaster horizontalis*). But most escapees do not get far. In rural districts, with dispersed habitation, it is common to see Ground Elder (*Aegopodium podagraria*) and Greater Celandine (*Chelidonium majus*) in the hedgerow banks next to cottages, having spread by seed from the cottage garden – but they rarely get further than 25–50m (80–160ft) away. Likewise, pink and white-coloured forms of wild Primrose (*Primula vulgaris*) may be found in similar locations, these being a result of pollen transfer by bumble bees between coloured Primula cultivars in cottage gardens and the wild type Primrose.

Mexican Fleabane (*Erigeron karvinskianus*), a non-native garden escapee found on wayside walls.

Broad-leaved Everlasting Pea (*Lathyrus latifolis*), a non-native garden escapee frequent on waysides and waste ground.

Our native Wild Bluebell (*Hyacinthoides non-scripta*) is similarly threatened by hybridisation with garden ornamental aliens. Cultivars of the garden non-native Spanish Bluebell (*Hyacinthoides hispanica*), available in a range of colour forms, are inter-fertile with, and hybridise freely with, our native Wild Bluebell. Both species are normally bumblebee pollinated. The foraging distance of a bumblebee is a few kilometres, and all wild native populations of

Colour hybrids between wild type and garden Primrose (*Primula vulgaris*) found on waysides close to habitation.

Bluebell in the British Isles are within bumblebee foraging distance of gardens where cultivars of alien Spanish Bluebell are present. Despite frequent hybridisation, and introgression of Spanish Bluebell genes into wild native populations, both hybrids and genes do not get far (Kohn et al., 2019). It is tough in the wild, and natural selection soon eliminates most of the garden ornamentals that venture out. In this example it appears that the Spanish Bluebell is ill adapted to our climate and the wild, and presents no more than a moderate local threat to our native species.

Ornamental Perennial and Annual Meadows

Ornamental perennial and annual meadows are communities of non-native ornamental perennial or annual garden flowers that are grown together in a random or designed manner to give the appearance of a meadow. The defining feature of both is that they are composed primarily of colourful alien ornamental species. These communities are not based on a community type that occurs naturally in Britain or Ireland. As the principal component species are alien, these ornamental meadows tend to be transient and non self-sustaining, especially if sown in landscaped settings or the wild. Nor do these artificial meadows perform well regarding the provision of ecosystem services. The growing of alien ornamental species in a meadow-like community does not enhance nature, or change the benefit or threat that aliens may present to ecosystems. It will always be better for nature if native species are used when creating or restoring meadows and other wild vegetation in the garden and park.

A municipal roadside meadow in Wales consisting of alien annual species.

The Wild Garden

The tidiest, most sterile and most artificial garden or park is usually almost totally devoid of wild life (but never completely). It is a constant battle keeping everything looking perfect and keeping nature out. But easing up on that battle and giving space to nature is good for the environment. Replacing non-native ornamental cultivars and species with wild native species is the best way to give space to nature: creating meadows, tussock grassland, annual communities, wetlands and woodland native flora. Prepare the ground by scarification or cultivation, sow the native wild-flower seed, and hand over control to nature. With a little weed control and the passage of time, native plant communities will become established and mature, and these in turn will help create and sustain healthy ecosystems buzzing with life.

The conventional cultivated garden planted with alien ornamental cultivars, and the wild garden consisting of native species, are qualitatively (and quantitatively) different. They are not the same. A wild native ecosystem cannot

A garden meadow consisting of native perennial species.

A native garden meadow with paths for access.

The astonishingly rare Lizard Orchid (*Himantoglossum hircinum*).

be gardened, otherwise it very quickly becomes a garden, and a perennial border of alien ornamentals cannot be left to nature, otherwise it quickly becomes a wild ecosystem dominated by native species. In the garden and park it is best to maintain a clear separation between the gardened areas and the wild areas, as each requires a completely different intensity of approach to management. In the gardened areas the gardener has dominant control of and responsibility for the plants, their condition, their arrangement and their aesthetic. In the wild areas nature has dominant control of the plants, their condition, their arrangement and their aesthetic. To intimately merge the cultivated garden and the wild garden, one with the other on the same spot, is to fudge the difference. The value of each is in their difference, and shared control between gardener and nature produces something that is neither completely wild nor completely cultivated.

Nature can surprise in response to the simplest of nature-friendly inactions. A friend of a friend near Frome, with the smallest of front lawns 2 × 3m (7 × 10ft), held off mowing for 'No Mow May', and up popped a Lizard Orchid (*Himantoglossum hircinum*), a Schedule 8 protected species and only the third recorded for all of Wiltshire. It is astonishing that the single tiny seed from which this magnificent plant must have grown, travelled so far unaided, and fell on friendly ground.

Wild gardening is an adventure. A native meadow is a good start, then add to it colourful patches of native annuals, perennial borders of interesting wild native species, nature ponds, rocky features, wild orchards, tussock patches and other habitats, all providing a welcome home for invertebrates, reptiles, amphibians, birds and mammals. It is a conceptual change in the way to garden. Nature ceases to be a threat. It is a welcome guest to be observed and enjoyed, whether it is an interesting weed that has popped up in the cracks of a path, dandelions and daisies in the lawn, a butterfly that has flitted by, or a wind-borne rare species that has landed on a friendly patch.

CREATING AN ORCHID MEADOW AT NEW MILL

David and Jean Trudgill

In 2003 we started to convert a small paddock that had previously been grazed by sheep, into a wild-flower meadow. We prepared the ground and sowed a mixture of grasses and local wild flowers. We also spread some seed of Lesser Butterfly Orchid that four years later, to our great surprise, resulted in a few plants appearing and flowering. That was the start – now, twenty years later, we have grown fifteen species of native orchids in our meadow, all from seed spread across it with the help of a breeze.

How has it been possible for us to grow fifteen orchid species in a small meadow of about 0.14ha? The key has been to create diversity. Several species grow in all parts of the meadow; many grow almost exclusively around a pond that we dug. The pond is fed from a neighbouring stream that sometimes has a pH of 8.0 or more. A further three orchid species grow in the shade below a 'pussy' willow and a couple of birch trees that we planted when starting the meadow. We incorporated large amounts of lime around one birch tree, and that is the only area where White Helleborine has appeared. The following are the fifteen species of native orchid that we have successfully introduced by seed into our garden meadow and around:

Green-winged Orchid
Early Marsh Orchid
Marsh Helleborine
White Helleborine
Lesser Butterfly Orchid
Early Purple Orchid
Southern Marsh Orchid
Common Twayblade
Common Spotted Orchid
Pyramidal Orchid
Greater Butterfly Orchid
Heath Spotted Orchid
Broad-leaved Helleborine
Northern Marsh Orchid
Heath Fragrant Orchid

Management in all its aspects is the key to creating a diverse and interesting meadow with the potential to help conserve wildlife, including insects and amphibians. For all our plant introductions (except the trees) we have used seed, spreading it widely so that each species can colonise the niche it favours. We have introduced seed of a wide diversity of plants, and in the spring our meadow is a carpet of flowers, the pussy willow hums with the sound of bees, and the pond is full of spawning frogs and toads. Some introductions do especially well – we now have more than 1,400 plants of Pyramidal Orchid. Others, such as Green-winged Orchid, have struggled. But it is worth the effort as our small meadow is a constant delight, full of interest throughout the year, and it has been an amazing learning experience. Each spring we open our meadow and garden under the Scottish Open Garden scheme, attracting lots of visitors and raising money for charity.

The garden meadow of David and Jean Trudgill.

City, Road and Rail

Developed land accounts for 8.7 per cent of land cover in England (Gov. stats, 2022). Our cities would benefit from more greening, to help purify polluted air, to help clean dirty water, to give a welcome in the city to nature, and to improve the well-being of city dwellers.

There is opportunity for additional green infrastructure on road networks, railways, waterways, industrial and retail parks, hotel and leisure complexes, all connected by urban green corridors and webs. Our cities would benefit from more trees, more urban meadows, more green roofs (Parkins and Clark, 2015) and more green façades. Nature's response will be surprisingly fast (Mata et al., 2023; Hoffman and Renner, 2020), and all the stronger if native species are used (Berthon et al., 2020).

Road and Rail Networks

Transport is a major source of pollution, and one that is difficult to avoid. It is also an important linear web-like network of natural verges, waysides and isolated spots, not intensively managed, and teaming with wildlife. New routes create an opportunity for habitat gain when they carve through intensively managed farmland. The greening of transport routes is helping to offset the harm often associated with development, and is making a huge contribution to nature recovery.

Canals and Waterways

Often the strip of land beside canals and waterways is unsuitable for development and intensive land use, and may provide an opportunity for creating corridors of natural habitat. Emergent native species at the water's edge can reduce flow rates, filter debris and stabilise banks, reducing erosion. The increasingly frequent river storm surges, often exacerbated by development and intensive agriculture upstream, may be moderated by restoring watercourse meanders and creating functional floodplain meadows using native species.

Retail, Industrial and Business Parks

Meadows are the ideal ground cover for landscaped zones associated with commercial premises, attracting and supporting wildlife while reducing maintenance costs. On poor to moderately fertile substrates one annual cut in autumn, and the cut removed, is often all that is required.

Hard landscaping and a mono culture of amenity grass in a new development.

Hard landscaping and a mono culture of amenity grass is high maintenance and provides very few ecosystem services.

Wasteland and Derelict Land

City wasteland and derelict land is surprisingly rich in wildlife, if not disturbed too often. They are fantastic and rewarding places to look for naturalised aliens, the mapping of which will give warning of the changes in the distribution of potential invasive aliens.

City Streets

Nature will come back to our city streets if we hold back from 'clean up' spraying of pavements and odd patches of hard standing with herbicide, and hold off mowing parks, amenity grassland and verges until the summer is over. It all

Cracks in paving providing a habitat for native species.

helps. Bare patches or plots beside walkways can be raked over and sown with native annuals and a meadow mixture that will help to clean air and water as it grows, and when it flowers, butterflies and bees will flock to it in surprising numbers. Pavement cracks and walls, if not sprayed, are rich hunting grounds for botanical curiosities.

Vertical Meadows and Green Roofs

Vertical meadows can be attached to the façades of buildings, and horizontal meadows may be placed on top, adding another dimension to city green infrastructure. Vertical meadows are created from native seeds embedded in a recycled supportive fleece medium that is attached to a supportive frame on the vertical sides of buildings, with drip irrigation provided. Horizontal meadows may be grown on substrates placed on flat or shallow sloping roofs. The choice of native species for meadows on buildings is huge, and will depend on the type of sowing medium, the situation and the location. Most situations on buildings will favour drought-tolerant, long-lived, rhizomatous species such as Wild Marjoram (*Origanum vulgare*), Lady's Bedstraw (*Galium verum*), and the wide range of native and naturalised Stonecrop species (*Sedum sp*) found growing in the wild.

ABOVE LEFT: Vertical meadows using native species give clean air and biodiversity in city developments.

ABOVE RIGHT: Shaded damp places within and between buildings can provide a habitat for native ferns, mosses and liverworts.

Under Cover

Native meadows may also be grown indoors where sufficient moisture and light can be provided, as in an atrium, a conservatory or a glasshouse. An indoor ecosystem can be a delight, bringing nature up close, and an indoor meadow will release refreshing oxygen and attract pollinators if a window is left open. Where light is a little dimmer but moisture plentiful, native ferns and mosses will flourish.

CHAPTER 7

SEED ORIGIN AND PROVENANCE

This chapter is about genetics, and is optional, if you wish. In it we discuss the choice of species and the origin of material (usually seed) that is used to restore nature, and the genetic considerations that are important in making these decisions. Although approaches differ in detail, worldwide, the consensus is that 'local origin native species are best' (Rivière et al., 2022; Hancock et al., 2020; Gamble et al., 2012; Bucharova et al., 2017; Kiehl et al., 2014). This is largely true for the British Isles, but there are a few exceptions. Also, every project is different and decisions on seed and plant origin will vary with project context.

Generally, the terms 'origin' and 'provenance' have different meanings when applied to forestry propagation material. The *origin* is the seed zone, or geographic location, often in the wild, from which tree propagation material (stock seed) was originally collected. The *provenance* is the geographic location of the tree nursery where the planting-out material was propagated. Therefore for trees, details of the origin, including characterisation of the original wild stand, can be informative of the climatic, genetic and ecological suitability of the propagated material. In contrast, the place of propagation (the provenance) will certainly have a bearing on the health and quality of the propagated stock, but the provenance is not informative of the climatic, genetic and ecological suitability of the propagated material. Thus 'origin' and 'provenance' might be defined as follows:

Origin: The place in the wild from which wild stock was originally collected.

Provenance: The place in cultivation at which wild stock was propagated.

This clear distinction in the application of the terms 'origin' and 'provenance' could beneficially be applied to seeds and plants of herbaceous restoration species.

The distinction between origin and provenance is important. For ecosystem restoration, origin will always be more informative than provenance.

Native or Alien Species

Wells *et al* (1981), in the first major publication on this subject, recommended using species that are native, common, widespread and attractive, for the creation of species-rich grasslands. Of course, the better suitability of native species compared with non-native species is intuitive and not surprising, as native species are those that have successfully colonised the British Isles in the current short post-glacial period by natural means (over the most recent 10,000 years), and have had opportunity to evolve adaptations to local conditions since then. Therefore, native species may be reasonably expected to be better adapted to conditions in the wild in the British Isles, and equally, it may be expected that non-native alien species are less well adapted to our conditions. From a native British flora of around 1,400 species, in Chapter 11 and the Appendix we list 195 native species as having potential for use in ecosystem restoration and creation, including familiar favourites such as Cowslip (*Primula veris*), Bee Orchid (*Ophrys apifera*) and Moon Daisy (*Leucanthemum vulgare*).

A number of archaeophyte species are also common, widespread, attractive and well adapted to the wild, and contribute positively to ecosystem services. Although not strictly native, some of these archaeophytes are familiar and important constituents of the wild. In Chapter 10 we list 22 archaeophyte species as suitable for restoration projects, including common species such as Corn Poppy (*Papaver rhoeas*), White Campion (*Silene latifolia*) and Common Mallow (*Malva sylvestris*).

There are around 1,500 naturalised neophytes recorded in the wild in the British Isles as self-sustaining and reproducing populations. These are mostly garden throw-outs or escapees, and almost all are best avoided because we cannot be certain that they are not aggressively invasive, will not disrupt mutualistic associations between native species, and are not harmful in other ways. But there are four exceptions, four naturalised neophytes that are common, widespread, attractive and well adapted, and appear to make a positive contribution to native ecosystems. These are Snake's-head Fritillary (*Fritillaria meleagris*), Hedgerow Cranesbill (*Geranium pyrenaicum*), Common Field Speedwell (*Veronica persica*) and Pineappleweed (*Matricaria discoidea*). These four naturalised alien neophytes are now welcome constituents of our wild flora.

The many thousands of non-naturalised neophyte (alien) species present in the British Isles are not suitable for use in ecosystem restoration. Almost all would not survive as self-sustaining populations in the wild, and some may indeed be harmful to nature if they did.

Wild Type or Cultivar

Modern cultivars are the product of twentieth-century plant breeding, and they have completely replaced the land races that were formerly grown in agricultural and horticultural crop production. Those land races, many of which are now lost, varied from locality to locality, having adapted to local conditions, and they had the capacity to continue to evolve in response to changing environments. Their demise came with the introduction of the EU Seeds Regulations. These regulations outlawed the trade in any cultivars not on the National List/Common Catalogue, and any seed of these cultivars that is not certified. The loss of the diversity of land races exposes the agricultural and horticultural sectors to controlled markets for seed and a much reduced gene pool of cultivars to choose from. As Dr Melaku Worede observe: 'If we all hang on to one string, the string will break' (Worede, 2009).

Agricultural land races of some native legumes and grasses have been grown for fodder in the British Isles since at least the 1620s. Mixtures containing (variously) Clover, Sainfoin, Lucerne, Sweet Vernal-grass, Meadow Foxtail, Meadow Fescue, Creeping Bent, Crested Dog's-tail, Cocksfoot, Timothy, Yellow Oat, Sheep's Fescue and Rough-stalked Meadow Grass were commonly in use for sowing pasture and meadow up until the start of the twentieth century.

All modern cultivars must be *stable* (breed true), *uniform* (not variable in form), and *distinct* (different to all other cultivars) to qualify as a cultivar and to be able to register for plant variety rights (PVR). Both land races of crops and wild types of native species are unable to meet the stable, uniform and distinct requirements. For clonal crops, such as those of Potato (*Solanum tuberosum*) and Garden Strawberry (*Fragaria × ananassa*), the requirements for stability and uniformity are automatically met because clonal cultivars are vegetatively propagated. However, the cultivars of the native species that are of interest to us, and that are subject to the Seeds Regulations, all reproduce by open-pollinated seeds. Therefore, to breed an agricultural, amenity or ornamental cultivar of a native seed species, and to meet the requirement for stability, uniformity and distinctness, it is necessary to breed out all the genetic variation in the stock, and to breed in all those desired novel features required of a cultivar. Consequently, the genetic mean and variance of the wild type is fundamentally different, qualitatively and quantitatively, to that of the corresponding cultivar. 'Wild type' and 'cultivar' might be defined as follows:

Wild type: Populations with a large genetic variance and potentially floating genetic mean.

Cultivar: Populations with a zero genetic variance and a fixed genetic mean.

Modern cultivars are genetically uniform and are unable to respond to environmental change in space and time (other than by returning to the breeder's field). The ability to respond to environmental change is a vital feature of ecosystem function, and therefore cultivars are a poor substitute for the wild type. For these reasons the modern cultivar forms of forage species, and the modern horticultural and ornamental cultivars of other native species are a very poor second best to the wild type when sowing a meadow and restoring nature.

British or Alien Ecotype

In the two decades from 1980, seed of native British species, but of wild mainland European origin, was imported into the British Isles at low cost and in appreciable quantities to meet the growing home demand for wild-flower seeds. These mainland European ecotypes, sown in the wild here, were instantly recognisable to experienced field botanists, and were rightly considered a threat to the scientific and heritage value of the native British and Irish flora (Monkswood, 2005). Since 2000, home production of native seeds has increased considerably and has almost completely replaced the importation of wild origin seed from mainland Europe. The import and export of wild seed across the UK/EU border is further reduced by the the recent introduction of phytosanitary regulations, and associated inspections and paperwork, post Brexit.

Local or Non-local Ecotype

'Local' may be defined as a zone described by the boundaries of an area from which source seed may have originated and within which it may be used. The smaller that zone, the more local will be the sourcing. The zone may be the British Isles (to which all native species will be local by definition), or countries within the British Isles, or a phyto-geographic region, an administrative region, or a nearby locality, or an immediately adjacent location, and the sourcing within that zone may combine samples from across the zone or be restricted to a population or niche within the zone.

Most native species exist in the wild as distinct ecotypes. For example, many species on Gower Peninsula are represented by large-flowered and large-seeded ecotypes compared to their inland counterpart. Ecotypes are populations within species that differ one from another across the natural range of the species in response to prevailing ecological conditions. Extreme variant populations may be found in extreme habitats, such as ecotypes with tolerance to heavy metal ions (Smith and Bradshaw, 1979).

The genetic differences between ecotypes may be demonstrated using 'common garden experiments', and for over 100 years these have been employed to identify and characterise genetic ecotypes, and have often demonstrated a *home site advantage* (Turesson, 1922; Joshi et al., 2001; Villemereuil et al., 2015). Nothing to do with gardens, these are experiments in which putative ecotypes from different *away* locations are grown together in randomised, replicated plots (or pots) at a common experimental site. Ecotypes observed at their home site will exhibit distinct phenotypic features, which, when they are grown in the common garden, can be teased out into genetic, epigenetic and environmental components.

The genetic component is that fraction of the variation between ecotypes (and within an ecotype) that persists in the common garden, and is heritable. The epigenetic component is that which may persist in the vegetative phase, but does not normally persist and is not normally heritable through the seed line. The environmental component is the plastic response of the genotype to the home environment, and this response normally fails to persist in the common garden.

Common garden experiment comparing different wild collections of Bird's-foot Trefoil (*Lotus corniculatus*) (Mortimer, 2016).

Most wild species have been shown to have heritable ecotypes. Wild Carrot (*Daucus carota*) is a good example. Inland ecotypes generally have a tall, erect habit at flowering, while coastal ecotypes can have a short, spreading habit at flowering. These differences are genetic. However, the situation can be complicated by *plastic* (non-heritable) responses to grazing. If the main tall inflorescence of inland Wild Carrot is grazed at flower bud, the plant often responds by producing exclusively ground-hugging spreading inflorescences.

It is attractive to speculate that the genetic differences between ecotypes, as revealed by common garden experiments, are indicative of a *home site advantage*. Home site advantage can be tested by performing a reciprocal transplant experiment (Johnson et al., 2021). In these experiments, all the ecotypes under study are grown in randomised replicated trials at the home site of each, and also at the away sites of all the other ecotypes being compared. Analysis of the results will often reveal a home site advantage in performance, when comparing performance at home and away sites, meaning that ecotypes tend to be better adapted to their home ground than elsewhere.

Having established that a home site advantage exists in nature it is tempting to assume that local is best when sourcing ecotypes for restoration projects. 'Local' is a measure of the distance between the source site and the restoration site, and that distance may be measured in geographic units, ecological units and genetic units.

Geographic Distance

The geographic distance between different ecotypes of the same species is often very small. For example, the phenology of flowering time is very responsive to natural selection (Voller et al., 2017) and can reveal differences between ecotypes over very short distances, despite free gene flow between those ecotypes. Common Knapweed (*Centaurea nigra*) growing in a meadow will often have evolved as an early-flowering ecotype in response to mowing, while the ecotype on the uncut field edge will often be a late-flowering ecotype. Studies of the Rothamsted Grass Park Experiment, established in 1856, have revealed that adjacent plots managed differently have, through natural selection, generated different ecotypes of Sweet Vernal Grass (*Anthoxanthum odoratum*) when separated one from another by only a few metres (Silvertown et al., 2006). In both cases, natural selection has proved a greater force than gene flow, and has maintained the integrity of these ecotypes over very short distances, despite unencumbered gene flow.

Local sourcing distances have never been satisfactorily defined in geographic terms, and the difficulty has been in making sufficient allowance for the fine grain distribution of ecological niches and ecoclines, and their associated ecotypes. Consequently, a number of different sourcing protocols exist, ranging from obtaining source material only from adjacent communities, to sourcing from within large administrative or phyto-geographical zones (Bucharova et al., 2017).

Ecological Distance

Many studies have found that ecological distance is a better measure of how well ecotypes have adapted than is the geographic distance between source and restoration sites (Smith et al., 2007; Raabova et al., 2007; Mortimer, 2016). The shorter the ecological distance between source origin and restoration site, the better the ecotype will have adapted for the intended use. This is intuitive, and in the absence of seed sources from adjacent wild communities, it is probably better to base sourcing policy primarily on phyto-ecological zones than on phyto-geographical zones.

Genetic Distance

The genetic distance between populations on source sites and restoration sites may be difficult to determine, but greater genetic relatedness between the two may be expected to yield greater success of restoration (Montalvo and Ellstrand, 2000).

Ecotypes are a natural selection response to specific sets of environmental conditions. They are ecologically adapted to prevailing conditions, but not at a 'local' geographic scale as we may understand it, but rather at the scale of the pattern of patchiness of the prevailing ecological variables. Consequently, the natural distribution of ecotypes mirrors the distribution of ecological patches and gradients.

It is intuitive, and most often true, that a very local wild source for stock of a native species is most likely to consist of ecotypes that are well adapted to the restoration site (Breed et al., 2013). Therefore, in sensitive natural and semi-natural wild landscapes, and within seed dispersal range and pollen dispersal range of surviving wild native plant ecosystems, it is wise to use well-adapted local origin ecotypes for restoration. The reason for this is to ensure that restoration is successful and beneficial to the wild, and does not threaten existing ecosystems. In completely degraded landscapes native species are still important, but local sourcing is less important. High genetic variance becomes of greater importance than local adaptation, and therefore a mixture of ecotypes of diverse origin may be used. An admixture of diverse origin ecotypes may best provide the high genetic variance that will enable restored populations on highly degraded sites to respond to diverse site conditions and changes in environment over time.

When Local Is Best

Local is best in the following circumstances:

- When restoring wild vegetation adjacent to existing high-quality wild vegetation. Here the best option is brush-harvested seed from the adjacent site.
- When restoring wild vegetation within distinctive phyto-geographic regions, such as the Gower Peninsular in Wales or the Isle of Wight in England.
- As a precautionary fallback position.

When Local Is Not Best

Local is not best in the following circumstances:

- When the local ecotype is a survivor of a historic adaptation to an environment that no longer exists due to environmental change (Jones, 2013).
- When the local ecotype is represented by an extreme adaptation with a very low genetic variance.
- When the species of interest does not exhibit ecotypic variation (Houston and Wolff, 2012).
- When the restoration site is highly degraded, large, heterogeneous, unstable and subject to climate change uncertainty. On such sites, mixed origins with high genetic variance may be expected to survive better over time and contribute more to ecosystem services (Havens et al., 2015).

In conclusion, it may be expected that the best adapted source material will be obtained when there is a close ecological match between the donor and recipient site, when the geographic distance between these sites is less, and when the genetic distance between these sites is also less. Source material, selected in this way, will usually be better adapted to the restoration site than material of non-local origin. However, well-adapted ecotypes may be low in genetic variance, and on some large disturbed sites genotypic admixtures derived from different non-local origins may better provide the genetic variation needed to allow novel ecotypes to evolve in what may be an uncertain environment (St Clair et al., 2022; Bucharova et al., 2018).

Genetic Benefits and Threats

Genetically robust populations, species, communities and ecosystems are those that are best able, and sufficiently fit, to exist and reproduce sustainably in the particular environment that they may call home. Even more important for sustainable survival is the genetic ability to respond, by natural selection, to environmental heterogeneity over space, and environmental change over time (Rogers et al., 2004; Monkswood, 2005). As we will see, local ecotypes that are extremely well adapted, and the adaptability that flows from a high genetic variance, are opposite sides of the same coin. We may want both in full measure, but we cannot have both without compromise.

Genetic Mean and Variance

Ecotypes can be described by the genetic means and the genetic variances of their traits, as described in Chapter 2. Generally, the more that an ecotype is well adapted, the more extreme will be the mean of the traits that define the ecotype, and the less will be the genetic variance of those traits. As the genetic mean of an ecotype responds by natural selection to the environment, it shifts to the extreme of the normal range for each trait, and it does so by losing genetic variance at the other end of the range. Extreme ecotypes are more frequent at the edge of the normally occupied territory and habitats for the species. The apparent extreme well adaptedness of these edge-of-community ecotypes comes at a cost to the genetic variance, which is often considerable. In restoration, the genetic variance is more important than the genetic mean. Therefore, in sourcing ecotypes for restoration it is often better to obtain mid-range ecotypes with high genetic variance, taking these from the mid-range of distribution in the wild, rather than from the edge of range. Ultimately, high genetic variance, rather than ecotype means, drives healthy and fit ecosystems.

It is tempting to visualise the variety and mobility of genes in a healthy ecosystem as a minestrone soup:

> Life is a minestrone
> Served up with Parmesan cheese
> Death is a cold lasagne
> Suspended in deep freeze
> Love is a fire of flaming brandy
> Upon a crêpe suzette
> Let's get this romance cooking, honey
> But let us not forget
> Life is a minestrone
>
> 10cc 1975

The quantity and quality of the genetic variation in an ecosystem will change over time. It may decrease in response to inbreeding, isolation, mortality, habitat destruction and catastrophic environmental change, all resulting in population decline and biodiversity loss. On the other hand, genetic variation may increase as a result of migration, hybridisation, introgression, spontaneous mutation, good connectivity, high gene flow, population growth, and ecosystem restoration/creation.

Population Size

Native species of the British Isles have suffered huge reductions in population size over the last 100 years due to agricultural intensification and development of the built environment. Population size is important for native species because it represents the stock of genetic material that is the fuel of natural selection and evolution. Most native plant species are partial or obligate outbreeders, requiring a minimum of two compatible individuals within pollination range to produce viable seed – for example, Cowslip (*Primula veris*) and Red Campion (*Silene dioica*). Outbreeding species drive evolution, providing for the future well-being of nature, and to do so they require large interbreeding populations.

Interestingly, some native species have temporarily, partially or completely, opted out of evolution, having locked into what, for them at a point in time, may appear to be a winning genetic formula by adopting an apomictic, selfing or vegetative lifestyle. For these species, population size is a little less critical for survival. The apomictic microspecies of, for example, Dandelions (*Taraxacum sp*) and Hawkweeds (*Hieracium sp*) require only one individual to produce viable seed, as seed develops somatically without pollination. Barring spontaneous mutation, all apomictic seeds from one plant are genetically identical. Self-compatible species, such as Bee Orchid (*Ophrys apifera*) and Tomato (*Lycopersicon esculentum*), also require only one individual to produce viable seed. In these self-pollinating species the pollen and the ovules are produced by meiosis, and the seed is formed by zygosis of the two, a process that can reduce genetic variation within individuals at a rate of up to 50 per cent per generation. Nevertheless, a single flower spike of Bee Orchid can produce almost a million seeds.

For all wild ecosystems, and the populations of native species on which they depend, bigger is better.

Population Isolation

Populations within species, defined as a cluster of individuals connected by interbreeding networks, may be isolated one from another by phenology, geographic distance, physical barriers, lack of pollinators and incompatibility between ecotypes.

Small isolated populations of outbreeding species are at risk of inbreeding depression and extinction. This can be remedied by increasing the population size and the genetic variance, and/or by connecting up with other populations of the same species by creating habitat stepping-stones and corridors, thereby increasing the *effective* population size. Connected is better.

Genetic Drift

Genetic drift is postulated to occur in small and declining populations, although hard evidence is scarce. The theory proposes that in outbreeding and inbreeding species the random assortment of genes from a small gene pool, at breeding, may lead to some random loss and some random retention of genes.

However, all genes that are expressed in the phenotype will be subject to the rigours of natural selection in the wild, and any change in the population genetics of these genes will not be random drift. But if genetic drift is seen as a threat to restoration it can be countered by increasing population size, the genetic variance and gene flow.

Gene Flow

Gene flow across the landscape, long and short distance, is mediated by pollen and seed dispersal in out-crossing species, and by seed dispersal alone in apomictic and obligate selfing species. Although native herbaceous vegetation is firmly rooted to the ground, its genes are mobile. Most wind-dispersed seed and pollen will fall to the ground beneath and near to the mother plant, but the tail cast by the seed or pollen shadow may be very long. For example, the dust-like seed of Lizard Orchid (*Himantoglossum hircinum*) is occasionally carried on the breeze hundreds of kilometres from northern France to Kent.

Gene flow between populations is a good thing, introducing refreshing and novel genetic material that will be sifted by natural selection for beneficial genes, while the non-advantageous are discarded. Using native species to create habitat corridors and stepping-stones across the landscape, linking populations, communities and ecosystems, will create pollen and seed dispersal routes, and will assist gene flow.

Genetic Swamping

It may be feared, and is postulated, that introducing large numbers of a non-local ecotype of a native species into the range of a local ecotype may swamp the local ecotype. There is no evidence of this having happened with ecotypes of native restoration species. However, there is good evidence of genetic swamping of wild ecotypes of Perennial Ryegrass (*Lolium perenne*) and Red Fescue (*Festuca rubra*) by the very widely sown agricultural and amenity cultivars of these two species. This is unlikely to happen with restoration seeding using native stock of wild origin, as the quantities involved are very modest compared with agricultural and amenity seeding. However, small populations of sensitive species represented by extreme ecotypes with poor fitness may be vulnerable to genetic swamping and should be safeguarded accordingly. Increasing the availability of adjacent favourable habitat, and taking measures to increase the population size of the vulnerable ecotype, may achieve this safeguarding.

Introgression

Introgression is the process by which one favourable gene, or a cluster of favourable genes, pass naturally from one ecotype to another, or from one closely related species to another. The process starts with a natural hybridisation event (F1) and is followed by repeated back-crossing (B1, B2, B3 and so on) to the resident ecotype or species, with natural selection ensuring that the favourable gene, or cluster of genes, passes down the line at each generation. The culmination is the incorporation of the favourable introgressed gene or cluster of genes into the population of the resident ecotype or species. Introgression is an important evolutionary process contributing to the genetic well-being of ecosystems, and introgression may be beneficially enabled by incorporating a targeted diversity of ecotypes and species into restoration projects.

Heterosis

Heterosis is the hybrid vigour observed when two different ecotypes or species hybridise naturally, increasing loci heterozygosity in the offspring compared to the parents. Hybridisation is more frequent in some genera than in others, and is often an indicator of weak barriers to crossing between related species, and a sign of speciation in action. Common Spotted Orchid (*Dactylorhiza fuchsii*) and Southern Marsh Orchid (*Dactylorhiza praetermissa*) readily hybridise, often producing vigorous F1 hybrids. These hybrids are fertile and can frequently interbreed to produce a persistent and vigorous hybrid swarm consisting of inter-crosses between generations of hybrids and back-crosses to both parents.

Hybridisation is not an option for apomictic species, for example Dandelions (*Taraxacum sp*) and Hawkweeds (*Hieracium sp*), because apomicts are non-sexual. However, the success of these species probably derives from them having locked in heterosis at conversion to apomixis.

Hybridisation and heterosis are important genetic features of a healthy and vibrant restored ecosystem, and may be encouraged by bringing together species and ecotypes that freely mate. Consider sowing genotypic admixtures on disturbed and unstable sites (Li et al., 2017).

Inbreeding Depression

Inbreeding depression is the opposite of hybrid vigour and results from the selfing or the crossing of two individuals that are too closely related. Increasing degrees of inbreeding bring together increasing numbers of lethal and sub-lethal double recessive genes, and as a result fitness nosedives. Small isolated populations are exposed to inbreeding depression. At below fifty to sixty individuals, the population can be sucked into the extinction vortex. Inbreeding reduces fitness, which reduces population size, which increases inbreeding, which reduces fitness further, and so on, until unavoidably the population dies out. To avoid the extinction vortex, or to rescue a population from the extinction vortex, increase population size, gene flow and the genetic variance (Johnson et al., 2010; Lobo et al., 2021).

Outbreeding Depression

Outbreeding depression is the partial or complete failure of crosses between parents that are too distantly related (Montalvo and Ellstrand, 2001). It is nature's way of maintaining the integrity of advantageous gene clusters, ecotypes and species.

CHAPTER 8

SURVEYING, PLANNING AND METHODS

Having decided to create or restore a species-rich wild meadow, or another type of wild herbaceous native vegetation, the first practical step is to survey the creation/restoration site. This provides a base line that can be compared with subsequent surveys, and provide information that will inform decisions on species choice, seed origin choice, ground preparation, sowing, early years' and long-term management. We also look at options for natural regeneration.

Botanical and Feature Surveys

The chosen restoration/creation site may be a window-box, a street planter, a patch of lawn, a park, some wasteland, a field, a farm, or a phytogeographic region. It could be any spot, anywhere. Every small or large patch lacking nature has potential. A botanical survey and a features survey will identify the existing ground cover and inform decisions on what may be retained, improved or removed.

Botanical Survey

If it is a large site of, for example, several fields, divide the site into enclosures or blocks and survey each separately. A botanical survey may consist of a walk over of each enclosure, taking a random/spaced route, and recording species and communities as they are encountered, noting if the vegetation is 'unimproved' (completely natural, species rich and long established), 'semi-improved' (species poor), or 'improved' (completely artificial), and noting if the species found occur throughout the enclosure, or are confined to the margins. The best time to record is summer, and recording will be more complete if it is also repeated in spring and autumn, catching very early and very late flowering species. However, an experienced surveyor will be able to identify most species in mid-winter from vegetative parts and remains.

Making a visual assessment and assigning to each species a score on the DAFOR scale is a convenient way to most easily record abundance of species.

The botanical survey will reveal if the vegetation on the site is completely natural and long established, and if any special native species are present. In this case the recommendation would most often be to maintain the vegetation in its current condition by beneficial management and not to introduce further species. If the vegetation is found to be semi-improved native and species poor, the recommendation would most often be to restore species richness by over-seeding. If the vegetation is found to be completely artificial, with native species absent, or if the site is

DAFOR SCALE

- D = Species DOMINANT
- A = Species ABUNDANT
- F = Species FREQUENT
- O = Species OCCASIONAL
- R = Species RARE

FEATURES TO LOOK OUT FOR

- Existing and relic boundary features such as hedges, stone walls and ditches
- Ant hills
- Ridge and furrow
- Cultivation lynchets and livestock terracettes
- Mineral workings and industrial relics
- Veteran trees
- Scrub
- Springs and wet patches
- Historic built structures
- Historic routes
- Habitation ruins and relics
- Watercourses, ditches, ponds, canals and open water
- Scree, exposed rock and exposed soils
- In-field and in-hedgerow pollards
- Mobile dunes and inland unstable sands
- Semi-natural ancient and secondary woodland
- Scheduled ancient monuments and battlefields

unvegetated, the recommendation is likely to be to create species-rich vegetation by sowing an appropriate brush-harvested mixture or a site-specific selection of restoration species.

Feature Survey

Most sites will have features that may be worthy of retention and protection, and these may be incorporated into plans for habitat restoration and creation. See the box on the right for examples.

Ant hills and scrub.

Soil Type, Nutrient Status, Topography, Weeds and Stress

Soil Type

The mineral component and the organic component of the soil will together influence the soil fertility, pH, water-retention capacity, friability, and toxicity, if any. This in turn will determine the species and communities that the soil is able to support. If in doubt, take a spade and dig a hole exposing the soil profile. The presence of chalk or limestone below shallow topsoil will indicate a well-drained soil with moderate to poor nutrient status and a high pH. The presence of deep loamy clay will indicate a moisture-retentive soil with moderate to high nutrient status and neutral pH. The presence of sand with little topsoil will indicate a very well-drained soil with poor nutrient status and possibly an acid pH.

Soil profile with shallow infertile topsoil overlying calcareous subsoil.

Early morning buttercups in a Somerset meadow.

Nutrient Status

An analysis of a soil sample can provide, amongst other things, a measure of the *available* mineral nutrients, the percentage organic carbon content, and the pH. The fertility of the soil and the amount of plant growth it can support will depend primarily on the content of *available* nitrogen and phosphorus. Low values for N and P are better for establishing species-rich native vegetation. However, as a general rule, soils in the British Isles that are infertile or moderately fertile are suitable for creating and restoring native vegetation. Only the most extremely fertile soils should be avoided. Examples would be former sewage works and muck heaps.

When creating a meadow on bare soil, nutrients are progressively locked up in biomass down through the soil profile as the sown vegetation develops, and available nutrients decline sharply over the formative years of a meadow. This is visible as initial lush growth, followed in a few years by more moderate growth. For this reason, a soil analysis of cultivated soil may not always be a good guide to the fertility of vegetated soil and its suitability for establishing native vegetation.

Topography

Meadows are best mown each year and are therefore best sited on flat, level ground that is easy to manage with machinery. Ground that is too steep or rough for machinery may be managed as pasture or left unmanaged. Usually, the steeper the ground, the thinner the soil, and the greater the species diversity this may support. Aspect is also important, with south-facing slopes hotter and more drought prone than north-facing slopes.

Weeds

Generally, when establishing a meadow, the richer the soil and the greater the perennial weed burden, the

Drought-prone soils can exert stress on meadows, favouring deep-rooting wild flowers at the expense of shallow-rooting grasses and increasing species diversity.

greater will be the management required. Most annual weeds perform an important role in nursing the establishing sown perennial species, and are not of concern. A site survey should look out for potentially troublesome perennial and biennial weeds, as rhizomes in the ground or as seed in the soil seed bank. Commonly encountered, and indicative of post-establishment management needs, are the Weeds Act species Creeping Thistle (*Cirsium arvense*), Spear Thistle (*Cirsium vulgare*), Broad-leaved Dock (*Rumex obtusifolius*) and Curled Dock (*Rumex crispus*). Sites with high infestations of these species should be avoided unless it is possible, by repeated cultivation, to eliminate vegetative plants of these weeds and to deplete their soil seed bank.

Taking a very long view, the Weeds Act species are pioneer species that can dominate in the first five to ten years of establishing native vegetation, but will then die out naturally, provided they are not sustained by selective livestock grazing.

Stress and Competition

The survival and success of individual plants in a wild community will often be determined by their ability to tolerate environmental stress and/or their ability to compete with their neighbours. Different sites will present different stress and competition challenges. For example:

- A site consisting of well-drained exposed limestone will present a rather hostile substrate for most native species, favouring those that can tolerate high pH, water stress and nutrient stress, such as Salad Burnet (*Poterium sanguisorba*), Bird's-foot Trefoil (*Lotus corniculatus*) and Small Scabious (*Scabiosa columbaria*). There will be near-zero competition because individuals are mostly not in contact with each other in these high-stress environments.
- At the other extreme, a site consisting of a water-retentive fertile loam soil will be low on stress and will support near complete ground cover. Survival on a fertile soil such as this depends less on stress tolerance and more on competitive ability, favouring species that are able to compete with their neighbours, such as Common Knapweed (*Centaurea nigra*), Moon Daisy (*Leucanthemum vulgare*) and Hogweed (*Heracleum sphondylium*).

Taking account of on-site stress and competition allows better selection of restoration species.

Opportunities for Natural Regeneration

The site survey may also look at opportunities for natural regeneration. Natural regeneration is the method used to allow wild native plant communities to assemble naturally without the deliberate introduction of species. This is feasible if the desired native species remain present as viable seed in the soil seed bank, or if a good example of the desired community is situated adjacent to the restoration site, or if the desired species are mobile and may blow in on the wind as seed. Either way, community assembly will be slower than restoration by sowing brush-harvested seed from a good nearby wild donor site. There is a third natural regeneration outcome that can be just as rewarding, and that is to cease cultivating and tending a piece of ground and to quietly observe. In the old days this was called abandonment or set-aside; these days it is called wilding (Broughton et al., 2022; Ladouceur et al., 2023).

Defining Project Objectives

Having taken into account the botanical survey, features survey, soil type, nutrient status, topography, hydrology, weed burden and all other factors, the objectives may be defined in sufficient detail to allow for planning a programme of work and for sourcing of materials. All projects large and small can make a difference, and all can contribute to nature's well-being. The objectives, defined in broad terms, may include a description of the desired outcome at the end of the project, supported by the following:

- A proposed site plan
- The communities/ecosystems to create or restore
- The restoration species used and the sources for these species
- The methods to be used
- A programme of establishment work
- A programme for long-term management.

Seed-based Methods

Seed is best (Pedrini and Dixon, 2020; Ryan et al., 2008). Other methods (such as plugs, turf, transplants) are more expensive (Davies, 2002), less successful and less natural. After all, seeds are nature's primary chosen method for higher plant reproduction and dispersal, having evolved and diversified into a huge range of well-adapted seed forms over 350 million years of natural selection. The advantages of using seed are these:

- A native seed mixture can be brush harvested from a nearby donor site.
- Seed production of individual restoration species can be mechanised.
- Seed can be stored from season to season with minimal loss of viability.
- The sowing of seed on a restoration site can be mechanised.
- The distribution pattern of establishing species is a natural response to the site.
- The resultant community structure has a high natural component.
- The method can be applied cost effectively at very large scale.

Plant-based Methods

Plants of restoration species may be used alone, or in combination with seed, to restore and create meadows and other native herbaceous wild vegetation, or to introduce specific species that would otherwise be difficult to introduce as seed, or would take too long to reach maturity from seed. Plant raising, planting out on the restoration site and follow-up watering are all very labour intensive, limiting application of the method to small areas. Even on small areas it would be necessary to also sow seed to provide a wild matrix into which the raised plants could be inserted. The advantage of using plants is that they are quicker to reach flowering than are plants arising from seed, and they may be placed where they are required to be.

The plants used may be field dug from plants in cultivation (but never from the wild), pot raised, or raised as plugs in cellular trays. A degree of caution is wise as native plants derived from commercial ornamental horticultural sources may inadvertently carry invasive alien aquatic or terrestrial species, or may carry residual pesticides, unless it is specified otherwise by the supplier. Wild-flower turf is another option for small sites, with the turf laid directly onto prepared soil and watered until rooted in. Another method, specific to development sites, such as pipelines passing through sensitive vegetation or the construction of underground tanks, is to turf strip before excavation work starts, store the turf nearby, and then replace the turf on the back-filled soil, again watering until rooted in.

Restoration, Natural Regeneration and Wilding

There is considerable overlap between the concepts of restoration using seeds, natural regeneration (not using seeds) and wilding, in the context of the British Isles. All three concepts aim to provide the conditions necessary for the natural assembly of wild communities and ecosystems that are healthy, dynamic and sustainable, and delivering ecosystem services (Toit and Pettorelli, 2019). There are more similarities between these three concepts than there are differences, but the main distinguishing features of each, as they apply to us, are described in the following sections.

Betony (*Betonica officinalis*), raised in cellular trays and almost ready for planting out.

Restoration

Ecosystem restoration aims to create and restore species-rich native plant communities and dynamic ecosystems by introducing restoration species within the existing landscape structure of the enclosed lowlands and unenclosed uplands, employing existing and traditional management methods.

Natural Regeneration

Natural regeneration is the natural process that may, in favourable circumstances, create and restore dynamic ecosystems within the existing landscape structure of the enclosed lowlands and unenclosed uplands, either following abandonment or by continuation of existing and traditional management, but omitting the introduction of restoration species (Redhead et al., 2013; Broughton et al., 2022).

Wilding (Rewilding)

Wilding aims to restore dynamic ecosystems at a landscape scale, removing enclosures and allowing natural regeneration of ecosystems to proceed (eventually) to wood pasture, managed primarily by natural processes and wild mega fauna, all with minimal human intervention (Carver et al., 2021; Irwin, 2023; Schmitz et al., 2023).

There are often opportunities presented by projects to select and merge elements of all three of these important concepts. Decisions on this may be guided by the site conditions and project objectives.

Programme of Work

A programme of work helps to organise the timely completion of tasks. These tasks may include, in date sequence, the operations required to complete the project (Shaw et al., 2020). Some of the required tasks may include:

- Clearance of unwanted waste and clutter
- Completion of outstanding tree, hedge, scrub, watercourse and boundary management
- Pre-cultivation grazing, mowing and topping
- Securing the seed required for the project (Erickson and Halford, 2020)
- Cultivating or scarifying to create the required seedbed
- Broadcasting the seed on the surface
- Rolling in
- Monitoring results.

CHAPTER 9

GENERAL METHOD FOR MEADOW RESTORATION AND CREATION

In this chapter we describe the general method for restoration and creation, and discuss how Yellow Rattle and an annual nurse may assist in the establishment of wild vegetation.

General Method for Meadow Restoration

Many semi-improved agricultural grasslands retain a reasonable content of wild native grasses, having lost most wild-flower species following applications of broad-spectrum herbicides to control grassland weeds, and following the application of fertiliser to increase grass yields. The same is also often true of many long-established lawns and parkland amenity grass, sometimes having a surprisingly high content of mowing-tolerant native species. All these species-poor grasslands may be restored and diversified by over-sowing with seeds of a mixture of native restoration species. This method is now widely and successfully used in the British Isles to restore species-rich native meadow communities without the need to plough up semi-improved agricultural grassland, or long-established amenity grass (Pywell et al., 2007). The general method of restoration is as follows:

- Select species-poor semi-improved agricultural grassland, or long-established amenity grassland, on moderately fertile to very poor soil.
- Avoid grassland with high infestations of perennial weeds such as docks (*Rumex sp*), thistles (*Cirsium sp*), nettles (*Urtica dioica*) and bracken (*Pteridium aquilinum*).
- Cut and remove existing above-ground vegetation in mid-summer.
- Scarify hard-to-remove thatch, aiming to create around 50 per cent bare soil.
- Select a suitable brush-harvested mixture or a complete meadow-seed mixture consisting of native restoration species, including Yellow Rattle (*Rhinanthus minor*).
- Surface sow at 2–4g/m^2 from August to December.
- Roll in to give good seed/soil contact.

- Graze or mow to 31 March of the following year, to tread in the seed and to reduce competition from the grass.
- Cease grazing and mowing from 31 March and allow the meadow to grow and flower until mid-summer, then cut and remove.
- Aftermath graze or mow until 31 March of the following year.
- Subsequently repeat the cycle each year, allowing the meadow to grow and flower between 31 March and mid-summer.

The success of this restoration method is primarily dependent on the combined effects of scarification, autumn sowing and incorporating Yellow Rattle.

Scarification: To remove potentially phytotoxic thatch (Bonanomi et al., 2022) and to open up sites for germination (Kiss et al., 2020).

Autumn sowing: To allow vernalisation to break seed dormancy in Yellow Rattle.

Incorporating Yellow Rattle: To weaken and suppress coarse grasses.

General Method for Meadow Creation

Meadow creation is the method used to establish a meadow on prepared soil after removal of weeds and any existing vegetation. The technique is ideal for former arable land, temporary agricultural grass, recent amenity grass, cultivated areas in a garden, bare ground following engineering and construction work, and waste or disturbed ground – in fact, any ground on which the existing vegetation will not be retained.

- Choose a site on moderately fertile to very poor soil.
- Avoid ground with a legacy of high infestations by perennial weeds such as docks (*Rumex sp*), thistles (*Cirsium sp*), nettles (*Urtica dioica*) and bracken (*Pteridium aquilinum*).
- Cultivate the soil to remove weeds and other vegetation.
- Prepare a weed-free firm seedbed with a medium to fine tilth.
- Select a suitable brush-harvested mixture or complete meadow-seed mixture composed of native restoration species.
- Surface sow at 2–4g/m^2 at any time of year, as conditions allow.
- Include a nurse of cornfield annuals at an additional rate of 1–2g/m^2, if desired.
- Roll in to give good seed/soil contact.
- Allow the sown annuals, and the annuals from the soil seed bank, to grow and flower until the following mid-summer, then cut and remove.
- Aftermath graze or cut until 31 March of the following year.
- In this, the second year, allow the establishing sown perennial meadow species to grow and flower until the following mid-summer, then cut and remove.
- Subsequently repeat the cycle each year, allowing the meadow to grow and flower between 31 March and mid-summer.

The success of this meadow creation method is primarily dependent on having a weed-free, medium/fine seedbed, and rolling in after surface sowing. The time of year of sowing is not material and can be whenever conditions allow.

General Method for Other Habitats

The methods described above for meadow restoration and creation can equally be applied to other types of native herbaceous vegetation, with minor adjustment to suit different conditions and objectives, as described below.

Pasture

Pasture is grazed year round, therefore it is best to select restoration species that are grazing tolerant and have a tufted, rosette, creeping, stoloniferous or rhizomatous growth habit.

Sandy Soils

Sandy soils can be very drought prone and are best sown in the wetter months of autumn and winter.

Wetland and Water Edge

Floodplain meadows and other wetlands can be subject to seasonal and unpredictable inundation, and if this happens in the twelve months after seeding, a portion of the sown seed yet to germinate and anchor in will move with the flow of water; many wetland species have evolved adaptations to enable this method of seed dispersal.

Moor and Heath

The method for moor and heath is similar to that for meadows, except that some sites may not be favourable for using Yellow Rattle (*Rhinanthus minor*).

Wayside and Tussock

The establishment-phase operations are the same as those for meadows, but there will be limited or no allowance for mowing and grazing after establishment.

Hedgerow and Scrub

The establishment-phase operations are the same as those for meadows, and seeding can take place at the same time as planting a hedge or scrub, or indeed after, and there will be limited or no allowance for mowing and grazing after establishment.

Woodland

The method for the establishment of woodland ground flora is the same as that for meadows, except that the woodland must be established and the canopy closed, that any cultivation should be surface and shallow, and that there will be no grazing or mowing of the established vegetation.

Orchids

The general method for orchids is quite different to that for meadows. Select an existing species-rich patch of well established, short native grassland and sprinkle the seeds sparingly with a soil carrier on a still day. Rain will wash the seed into the soil profile.

Yellow Rattle (*Rhinanthus minor*)

Yellow Rattle is a common, attractive and widespread hemi-parasitic annual of species-rich hay meadows (Westbury, 2004). In the British Isles Yellow Rattle is the most common of a small group of hemi-parasitic grassland annuals that also includes Eyebrights (*Euphrasia sp*), Red Bartsia (*Odontites verna*), Field Cow-wheat (*Melampyrum arvense*) and Lousewort (*Pedicularis sylvatica*).

Yellow Rattle is green-leaved and obtains part of its nourishment from photosynthesis, but it also has an obligate requirement to parasitise certain of its grassland neighbours via root haustoria. Despite being green, Yellow Rattle will perish soon after seed germination unless it is able to make root contact with, and draw additional nourishment from, its nearby hosts. Host plants become visibly weak as a consequence and will often eventually die. Within a meadow community some species will be, to varying degrees, susceptible to Yellow Rattle parasitism, while other species will be, to varying degrees, tolerant of Yellow Rattle. We do not know the full host range of Yellow Rattle and we do not completely understand the basis for host susceptibility and tolerance. It is not that the haustoria are choosy, because Yellow Rattle will latch onto its own roots and to those of others of its own species.

The differential host range of different hemi-parasite grassland species, and the differences in the degree of susceptibility and tolerance of other species, provides an opportunity to use hemi-parasitic species as ecosystem engineers to assist in the restoration of native meadows, and as a tool to manipulate community composition and

Yellow Rattle (*Rhinanthus minor*) with companion species.

SOME SPECIES THOUGHT TO BE SUSCEPTIBLE TO OR TOLERANT OF YELLOW RATTLE (*RHINANTHUS MINOR*) AND EYEBRIGHTS (*EUPHRASIA SP*) (BROWN ET AL., 2021)

SUSCEPTIBLE to YELLOW RATTLE

Achillea millefolium (Yarrow) N
Centaurea nigra (Common Knapweed) N
Festuca sp (Fescues) N
Holcus lanatus (Yorkshire Fog) N
Hordeum secalinum (Meadow Barley) N
Lolium perenne (Perennial Ryegrass) N
Poa sp (Meadow Grasses) N
Trifolium sp (Clovers) N
Vicia cracca (Tufted Vetch) N

TOLERANT of YELLOW RATTLE

Anthoxanthum odoratum (Sweet Vernal Grass) N
Bromopsis erecta (Upright Brome)
Cirsium arvense (Creeping Thistle) N
Hypochaeris radicata (Cat's Ear) N
Leontodon hispidus (Rough Hawkbit) N
Leucanthemum vulgare (Moon Daisy) N
Lotus corniculatus (Bird's-foot Trefoil) N
Plantago lanceolata (Ribwort Plantain) N
Rumex acetosa (Sorrel) N

SUSCEPTIBLE to EYEBRIGHTS

Anthyllis vulneraria (Kidney Vetch) N
Cynosurus cristatus (Crested Dog's Tail) N
Festuca sp (Fescues) N
Lotus corniculatus (Bird's-foot Trefoil) N
Medicago lupulina (Black Medick) N
Plantago lanceolata (Ribwort Plantain) N
Poa sp (Meadow Grasses) N
Trifolium sp (Clovers) N

TOLERANT of EYEBRIGHTS

Centaurea nigra (Common Knapweed) N
Leontodon hispidus (Rough Hawkbit) N
Leucanthemum vulgare (Moon Daisy) N

N = Native

The crinkled first true leaves of establishing Yellow Rattle (*Rhinanthus minor*).

structure. Adding Yellow Rattle to existing improved and semi-improved agricultural grasslands usually results in reduced vigour and decline of coarse and competitive grasses such as Perennial Ryegrass (*Lolium perenne*), Yorkshire Fog (*Holcus lanatus*) and Rough-Stalked Meadow Grass (*Poa trivialis*). Combining Yellow Rattle seed with a complete mixture of selected restoration species and sowing this together into scarified species-poor grassland will ensure that competition from coarse grasses is reduced, giving the sown restoration species a better chance of survival. Also, Yellow Rattle, being an annual, dies after shedding seed in July, leaving myriad welcoming vacant niches for the sown restoration species to occupy (Pywell and Bullock, 2005; Westbury et al., 2006).

Yellow Rattle seed is golden in colour, quite large, roundish and flattened, with a narrow wing around its outer edge. The seed is notoriously short-lived and does not survive in the soil seed bank from year to year due to loss of viability and predation (Mudrak et al., 2014). Longevity of the seed can be improved by temperature- and humidity-controlled storage, allowing some viability to be retained from year to year. Germination is controlled by a need for vernalisation (while fully imbibed) to break dormancy. The requirement for imbibed chilling is absolute and considerable, and in the field, seed sown after Christmas will not usually receive sufficient natural chilling to break dormancy. The best time to sow to ensure maximum germination is to sow fresh new-season seed in late summer to late autumn. A spring sowing in the field will not germinate. The germination of viable vernalised seed occurs in waves from late February and through March. The seedlings are unmistakable, with crinkly first true leaves positioned between the smooth cotyledons. Yellow Rattle is not hardy, and early waves of germination can be blackened by frost.

In the field, introduced Yellow Rattle can multiply rapidly under good conditions, often booming in year two or three after sowing. Inevitably, high populations attract enemies (pests and diseases) and host depletion sets in, so a bust then follows the boom. The boom of Yellow Rattle typically gives way to patches dominated by Sweet Vernal Grass, Ribwort Plantain and Sorrel. These patches later give way to more diverse communities, and Rhinanthus host plants creep back in, as does Yellow Rattle, eventually. The boom/bust cycle in Yellow Rattle is around eight years. Within a field, different patches will be at different stages of the cycle, each experiencing its own localised Yellow Rattle cyclic succession. This drives increased community diversity over time.

Prepare the ground for Yellow Rattle by scarifying the chosen species-poor restoration grassland to create up to 50 per cent bare soil, and autumn surface-sow fresh Yellow Rattle seed at 0.2g/m^2 (range 0.01–0.5g/m^2) as part of a complete mixture of restoration species sown at 2–4g/m^2. Rolling in and winter grazing, if possible, will give good seed/soil contact.

Abundant Yellow Rattle can reduce the forage yield of grassland, sometimes dramatically, but control is easy. As the species is an annual with virtually no soil seed bank, one season of close mowing and/or grazing will eliminate Yellow Rattle completely from a field.

Yellow Rattle (*Rhinanthus minor*) in an established meadow.

Annual Nurse

Using the meadow creation method and sowing onto bare prepared soil, establishment of the sown restoration species is noticeably better if in the first year after sowing there is a good ground covering of annual species, either sown or arising from the soil seed bank. The annual species appear to nurse and protect the slower germinating and slower establishing perennial restoration species (Lopasio et al., 2021). Trials comparing removal of annual species with non-removal demonstrate that non-removal is of clear benefit to the establishment of perennial meadow species (these are personal observations). There are different ways of providing an annual nurse, as described in the sections below.

Cornfield Annuals

A mixture of native cornfield annual species (sown at 1–2g/m^2) is the ideal nurse, protecting the sown perennials, stabilising soils, providing pollinator services and being pleasingly colourful.

Cereals

A cereal, such as Oat (*Avena sativa*) or Barley (*Hordeum vulgare*), is an alternative. Cereals are quick to establish, will stabilise erosion-prone ground, will protect the sown restoration species, and will not persist beyond the establishment phase of a created meadow. However, cereals are wind pollinated and do not provide any pollinator services.

A cereal drilled as a nurse for slower germinating perennial meadow species.

Native and archaeophyte cornfield annuals sown as a nurse for sown perennial meadow species below.

Annuals Arising from the Soil Seed Bank

Most annual weeds arising from the soil seed bank will also play a beneficial role in nursing the sown restoration species, also providing pollinator services, and will not persist beyond the establishment phase of a created meadow. In this context, good weeds include Groundsel (*Senecio vulgaris*), Shepherd's Purse (*Capsella bursa-pastoris*), Scarlet Pimpernel (*Lysimachia arvensis*), Common Field Speedwell (*Veronica persica*), and many more. However, beware Blackgrass (*Alopecurus myosuroides*) because it can swamp new sowings, and beware Charlock (*Sinapis arvensis*) as it has phytotoxic capabilities inhibiting the establishment of other species within its range.

An annual nurse of native species arising naturally from the soil seed bank after broadcasting perennial meadow seeds.

Westerwolds Ryegrass

Westerwolds Ryegrass (*Lolium multiflorum*) was recommended as a nurse for meadow mixtures back in the 1980s, but experience has found it to be over-competitive, smothering, and persisting from year to year, preventing the proper establishment of a created meadow community. It differs from Perennial Rygrass (*Lolium perenne*) in having awned seeds. Westerwolds Ryegrass, also known as Italian Ryegrass, has become a serious agricultural and horticultural weed in recent times (Hill et al., 2022). Avoid it at all costs.

It follows from the above that if sowing a meadow in the autumn, winter or spring it is good to make provisions for an annual nurse cover, and good to hold off cutting this until mid-summer at the earliest, thereby gaining the greatest benefit from the annual cover. And how does this nurse effect work? It is probable that the annual vegetation protects the slower germinating seed of perennial species from predation, and the slower establishing seedlings of perennial species from extremes of temperature and moisture availability.

Westerwolds Ryegrass (*Lolium multiflorum*).

CHAPTER 10

RESTORATION SPECIES AND MIXTURE COMPOSITION

The restoration species are those species from our wild flora that may be considered as candidates for the creation and restoration of species-rich meadows and other herbaceous vegetation.

Restoration Species

Terry Wells recommended that the species used to restore and create species-rich wild vegetation should be native, common, widespread, attractive, and amenable to propagation and establishment from seed (Wells et al., 1981). In the foregoing chapters we have looked at those species and cultivars present in the British Isles that are not normally candidates for meadow and wild vegetation restoration and creation. We must therefore set aside the following:

- Schedule 8 protected species (69 of our rarest and most vulnerable native species)
- Sensitive species (around 1,000 rare, local and restricted native species)
- Schedule 9 prohibited invasive alien species (58 species)
- All non-naturalised alien species (around 80,000 species in gardens and collections)
- Almost all naturalised alien species (around 1,500 species and increasing)
- Most native and archaeophyte weedy and invasive species (around 50 species)
- All agricultural, horticultural and ornamental cultivars (around 50,000 cultivars).

We are left with a list of restoration species: 196 native species, twenty-two archaeophyte species, and four naturalised neophyte species. These species are listed in Table 2 (*see* Appendix), and are described individually in Chapter 17. The list is a guide and is not definitive. It may be adapted according to opinion, objectives, practicalities and local conditions.

Mixture Composition

Seeds of restoration species may be selected and blended in combinations producing a wide range of different meadow and wild vegetation communities. The core species (with sowing rates coloured blue in Table 2 (*see* Appendix)) are those that may be selected to make up a base community. The additional species (with sowing rates coloured green in Table 2 (*see* Appendix)) are those that may be added individually, if desired, to achieve specific objectives or to suit specific site conditions.

When combining the chosen species in a mixture, the proportions of each species may be based on the seed numbers per gram for that species. Species with large seeds will require a greater weight of seed, and those with small seeds a lesser amount, to achieve a certain density of established plants. Success of establishment will be determined by multiple interactions between species and seed lots, across sites, at different times of the year, and depending upon different weather and environmental conditions. Seed dormancy features may be considered when selecting species and quantities for sowing at different times of the year. Autumn sowing will favour species requiring vernalisation and a spring sowing will not. Yellow Rattle will germinate and establish from an autumn sowing, but not a spring sowing. Cowslip will also germinate and establish from an autumn sowing, but will lie dormant in the soil from a spring sowing, and because of good longevity, will germinate and establish in the following spring.

Experience, intuition and knowledge come into play when composing a seed mixture. Proportions based on seed number per gram is a good starting point, with an expectation of perhaps 10 per cent establishment from seed where the conditions are favourable, and perhaps 1 per cent survival in difficult conditions.

The average number of seeds per gram in a meadow mixture will be around 1,000. Therefore, taking this number, if the desired number of establishing individuals is, say, 100/m^2, the sowing rate of a complete meadow mixture will need to be between 1g/m^2 (at 10 per cent survival) and 10g/m^2 (at 1 per cent survival). Experience tells us that 2g/m^2 of a complete mixture is a good rate applicable to most typical ecosystem restoration projects in the wild, the semi-wild and on farmland, and that 4g/m^2 is a good rate in landscaping situations, parks and gardens. Do not exceed 4g/m^2 as this can lead to over-dominance of more vigorous species in the mixture.

Having selected the restoration species based on (1) suitability for the site conditions, (2) the species being a desired component of the resultant vegetation, (3) the species being within its native range, and (4) the availability of seed, the next stage is to determine the percentage content for each species in the mixture. This may be determined based upon (a) the target ground cover for the species, (b) the seeds per gram for that species, (c) the time of year of sowing, (d) the estimated percentage establishment, and (e) the anticipated interaction of the species with the environment and with other species over time. Proposed habitat seeding rates for species within mixtures are given in columns 3–13 of Table 2 (*see* Appendix). A sowing rate for a species of 0.05g/m^2 equates to 0.1 per cent by weight in a meadow mixture sown at 2g/m^2.

Annual Wild Flowers

Annual vegetation, either sown on, or arising from, the soil seed bank, supports an astonishing range and quantity of invertebrates in summer, often greater than that found in species-rich meadows in summer. Rare arable weeds such as Cornflower (*Centaurea cyanus*) and Corn Cockle (*Agrostemma githago*) were once common in farmers' fields, but are now rare in our modern herbicide-treated crops. The weed flora of arable cropping has changed in the twentieth century from colourful annual wild flowers, which are now easily killed with herbicides, to grass weeds such as Blackgrass (*Alopecurus myosuroides*) and Wild Oat (*Avena fatua*), which are not so easily killed with herbicides.

Volunteer Common Poppy (*Papaver rhoeas*) in a seed crop of Cornflower (*Centaurea cyanus*).

Annuals of cultivation arising from the soil seed bank.

Most of the annual wild-flower species formerly associated with arable cropping are colourful archaeophytes, introduced with the crops of early settlers, while others are native annual wild flowers of disturbed ground. Column (3) of Table 2 (*see* Appendix) lists six core restoration annual species and twenty-seven additional restoration annual species, for use in creating species-rich annual communities of colourful wild flowers. This is easily achieved by sowing a selected mixture of these species onto prepared soil in the autumn, winter or spring, and maintaining the ground from year to year by an annual surface cultivation after seed dispersal.

Neutral Lowland Meadow and Pasture

Most unimproved (agriculturally) lowland meadow and pasture on neutral soils can be classified as MG5 (mesotrophic grassland 5), or subsets of MG5. These are species-rich grasslands on neutral loam soils (pH 5.5–7.0), where the soil may consist of alluvium, clays and sands in varying proportions. Few MG5 meadows have been spared the plough, the sprayer or the fertilizer spreader.

Column (4) of Table 2 (*see* Appendix) lists twenty-three core species (in *blue*) and thirty-four additional restoration species (in *green*) for neutral meadow and pasture. Combinations of these species may be used to restore existing semi-improved (species-poor) grassland by over seeding, or to create species-rich lowland meadow and pasture by sowing onto bare prepared soil.

Calcareous Meadow and Pasture

Most unimproved (agriculturally) calcareous meadow and pasture can be classified as CG2, or subsets of CG2. These are species-rich grasslands on alkaline, well-drained soils over limestone or chalk (pH 7.0–8.5). The best calcareous grasslands are often on thinner soils on, for example, the North and South Downs, Purbeck, Salisbury Plain and the Cotswolds, in situations where they have been spared the plough, the sprayer and the fertiliser spreader.

The restoration species for calcareous meadow and pasture are listed in column (5) of Table 2 (*see* Appendix). The high species number reflects the characteristic high species diversity of calcareous grasslands. Selected combinations of these restoration species may be used to restore existing semi-improved (species-poor) grassland by over seeding, or to create species-rich calcareous, meadow and pasture by sowing onto bare prepared soil, often with considerable success.

Vegetation on Sandy Soils

Sandy soils are usually very well drained, drought prone, nutrient poor, easily disturbed, and usually neutral to rather acid in pH. They are therefore not the best land for farming, so are often not farmed, or if they are farmed, are used for extensive grazing or horticulture. Rarely are light sands in arable cultivation or intensive forage production, and would not normally produce enough growth to be managed as a meadow. Sandy soils will, however, support very species-rich, rather open, wild native vegetation. In the table of restoration species we identify twenty-three core species and thirty additional species for sandy soils. Selected combinations of these species may be used to diversify existing species-poor communities by over seeding, or to create species-rich communities by sowing onto bare prepared ground. Light sands are best sown in the autumn so that seedlings can establish in the wetter months. A spring sowing is at risk from possible summer drought.

Floodplain, Wetland and Water Edge

Column 7 of Table 2 (*see* Appendix) lists the restoration species that may be used to restore and create floodplain meadows, wet meadow and pasture, other wetlands, and the edge of freshwater ponds, lakes, ditches and watercourses. Floodplain meadow communities (MG4) are one of the major casualties of the plough and agricultural intensification because they were traditionally located on the fertile flat land beside rivers and streams. Only around 2,000ha (4,950 acres) survive in Britain.

MG4 floodplain meadow at Clattinger Farm SSSI.

The restoration and creation of floodplain meadows is a priority in central lowland England, to help improve river water quality and to store floodwater during storm surges. The sooner a start is made the better because it takes decades from seeding for nature to assemble the complex community structure characteristic of floodplain meadow MG4 grassland (Woodcock et al., 2011). The reason for this is the time taken to build stable communities in response to complex hydrology. Created and restored meadows on free-draining neutral and calcareous soils are quicker to reach maturity.

Upland Meadow and Pasture

The soils of upland meadow and pasture will often be rather acid and the meadows may not be as species rich as their lowland cousins. Column 8 of Table 2 (*see* Appendix) lists twenty-one core restoration species and thirteen additional species that may be suited to the restoration of upland meadow and pasture.

Welsh upland meadow.

Moor and Heath

Moor and heath, like all wild native vegetation in the British Isles, has suffered from agricultural improvement, particularly in the lowlands. Formerly, moor and heath was widespread on nutrient-poor, rather acid, sometimes wet soils, and was often managed by extensive grazing. Many species typical of moor and heath are not easy to grow in cultivation, and often the best seed for restoration and creation will be that obtained by brush harvesting wild donor sites. Column 9 of Table 2 (*see* Appendix) lists the restoration species for moor and heath.

Wayside and Tussock

Wayside and tussock herbaceous vegetation is generally unmanaged from year to year, except for occasional scrub control to halt the natural tendency for grassland to become woodland. Waysides are the vegetation ribbons that line roads, tracks, footpaths, canals and railways, and tussock is the native vegetation that develops on other unmanaged land. Both are a huge potential resource for nature and for delivering ecosystem services. Where wayside and wasteland is degraded and species poor, the methods of restoration and creation can be applied to increase species diversity. Column 10 of Table 2 (see Appendix) lists the restoration species that may be used to enrich and create unmanaged vegetation.

Wayside spring flowers on a Gloucestershire country lane.

Hedgerow and Scrub

Hedges are visually prominent linear features of enclosed landscapes, but less obvious is the small scale, high diversity, niche eco-cline from the centre of the hedge to the field edge. The same applies to scrub, with a distribution of niches that is patchy rather than linear, but no less diverse. This high niche diversity can generate and support high species diversity, providing opportunities to use quite a large number of restoration species for the restoration and creation of hedge and scrub ground flora. Column 11 of Table 2 (see Appendix) lists thirty core restoration species and thirty-seven additional species for hedge and scrub.

Woodland Ground Flora

The British Isles have a temperate climate, and most woodland here is composed of deciduous tree species. In closed canopy mature woodland there is not much to see on the ground during summer, autumn and winter. In summer, the canopy draws moisture from the soil and casts dense shade. Few plants can tolerate the shade and the drought.

Spring hedgerow flowers in Somerset.

Spring woodland flowers in Wiltshire.

Come spring, as temperatures rise, as days lengthen, and before the trees break into leaf, the woodland ground flora comes alive with wild flowers capturing sunlight streaming through the bare branches. Spring woodland wild flowers rush to burst into flower so that they may attract early emerging pollinators. Very few grasses are able to survive in deciduous woodland. It is the domain of spring flowers, ferns and lower plants. Column 12 of Table 2 (*see* Appendix) lists a modest eleven core restoration species and twenty-nine additional species for woodland ground flora.

Coastal

The coast remains one of the wildest and best places for recreation and botanising. Much of the coastline has escaped agricultural intensification and development because often the ground is too difficult for either. So the coast is an important spot for diverse surviving wild habitats including shingle banks, mobile dunes, dune slacks, mudflats, salt marsh, cliff face, scree and maritime turf, all on a range of different substrates (chalk, shale, limestone), and all differing in microclimate depending on aspect, exposure and so on. Native halophyte species are characteristic of maritime vegetation and are tolerant, to a degree, of inundation or salt spray. They are found closest to the shore and often have seeds that are adapted for dispersal on the current. Further back from the salt line halophyte native species give way to non-halophyte native species, all finding a home across a wide diversity of habitats and niches. Table 2, column 13 (*see* Appendix) lists a sample of restoration species that may be used to create and restore wild vegetation by the sea.

Richly vegetated shingle bank at Chesil Beach, Dorset.

Orchids

Native Orchids are rare and always a great pleasure to find in the wild. They have dust seed, and each flower on an inflorescence spike may produce up to 1,000 seeds. If the mother plant is allowed to reach maturity, the mature seed capsules split longitudinally, releasing the dust-like seed to float on the breeze. Creating or restoring a meadow may, with luck, create conditions favourable for the establishment of Orchids, and a seed passing by may drop onto your plot, however small.

Most Orchid species native to the British Isles are rare, protected, or highly sensitive for reasons of science or heritage. A few, however, are sufficiently widespread and frequent to be considered as restoration species, and these species may be added as seed when restoring or creating suitable vegetation. A list of potential restoration Orchids may include around eleven species, these being the more common and more widespread of the fifty-one or so species of Orchid native to Britain and Ireland. All Orchid species are somewhat restricted in habitat preference and distribution, and this will be an important consideration when selecting restoration species for a project, restricting usage to the existing native range of each.

RESTORATION ORCHID SPECIES

Anacamptis morio (Green-Winged Orchid) N
Anacamptis pyramidalis (Pyramidal Orchid) N
Dactylorhiza fuchsii (Common Spotted Orchid) N
Dactylorhiza praetermissa (Southern Marsh Orchid) N
Epipactis heleborine (Broad-leaved Helleborine) N
Epipactis palustris (Marsh Helleborine) N
Neottia ovata (Twayblade Orchid) N
Ophrys apifera (Bee Orchid) N
Orchis mascula (Early Purple Orchid) N
Platanthera chlorantha (Greater Butterfly Orchid) N
Spiranthes spiralis (Autumn Lady's Tresses) N

N = Native

General Purpose Mixture

It may be noticed from the foregoing that a small group of native species are common to a wide range of habitats. Using combinations of some of these species it is possible to create native plant communities composed of a moderate number of generalist species. The advantage of this is that the created vegetation is likely to succeed across a wide range of soil types and conditions. A further important advantage is that the structure of these non-complex created communities is likely to be rather open and receptive to natural colonisation by other native species from the soil seed bank, from seed borne on the wind from the nearby wild, and from seed carried in by roaming fauna (Pakeman et al., 2019). Different species occupy different niches in a community, and the fewer the species, the greater the number of vacant niches. The box below lists thirteen generalist native species that combine well to produce a simple general mixture of species.

An opposite and alternative approach is to devise combinations of as many species as possible in the expectation that some species will survive regardless of the site conditions. This approach can be interesting, but it is often difficult to predict the outcome.

GENERALIST RESTORATION SPECIES

Achillea millefolium (Yarrow) N
Centaurea nigra (Common Knapweed) N
Cynosurus cristatus (Crested Dog's Tail) N
Daucus carota (Wild Carrot) N
Festuca rubra (Red Fescue) N
Galium verum (Lady's Bedstraw) N
Leucanthemum vulgare (Moon Daisy) N
Lotus corniculatus (Bird's-Foot Trefoil) N
Plantago lanceolata (Ribwort Plantain) N
Primula veris (Cowslip) N
Prunella vulgaris (Selfheal) N
Ranunculus acris (Meadow Buttercup) N
Rhinanthus minor (Yellow Rattle) N

N = Native

CHAPTER 11

SOURCING SEEDS

Columella in Roman times described the use of hay-barn floor sweepings to restore grasslands (Lange, 1976), and throughout the ages hayseeds have been used by farmers to return arable to pasture and meadow, and to patch up poached fields and gateways. Today we re-purpose these methods to patch up nature and restore ecosystems.

Native seeds may be harvested directly from the wild by a variety of methods, or may be grown and harvested in cultivation using a range of techniques. Of all the different options, the best, in terms of ease of operation and naturalness of outcome, is to brush harvest local native seed from a nearby high-quality wild donor site, although brush harvesting is limited in the British Isles by the scarcity of good-quality donor sites following twentieth-century agricultural intensification. Other methods for sourcing seeds are also important and will have applications suited to differing circumstances and differing site conditions. Some of these methods are discussed below, including methods used to grow and harvest seeds of native species in cultivation.

Hand-collected Wild Seed

With the permission of the landowner or occupier, and within the law, hand collection of wild seed can be a good way of sourcing seed of certain species for use in restoration. These may be restoration species that are difficult to grow in cultivation, such as Wild Garlic (*Allium ursinum*), Cat's-ear (*Hypochaeris radicata*) and Yellow-flag Iris (*Iris pseudacorus*), or species that may be particularly abundant with seed on a nearby wild patch such as Dandelion (*Taraxacum officinale*), Agrimony (*Agrimonia eupatoria*) and Foxglove (*Digitalis purpurea*).

It is always important to restrict the amount of seed collected to a quantity that does not harm the source population. A safe precautionary limit may be to pick no more than 20 per cent of the seed produced in any year. Collecting seed from the wild for restoration projects has a net benefit as it extends the dispersal range of the species concerned. Keep an eye on the wild population, and at peak seed maturation pick into plastic buckets or paper sacks. At the end of the day spread the seed to dry on a sheet or tray, and when dry, clean by hand using kitchen sieves. The dry seed may then be placed in sealed containers and stored in a cool place or a fridge.

If collecting a mixture of species from a meadow by hand, this can be done on a small scale using a sweep net, or by scything a section, drying on a sheet in the open, and thrashing out the hayseeds. Hand-held vacuum collectors are not advised because of the harm they do to invertebrates.

For restoration Orchids, do not collect seed from the wild. However, seed may be collected from land in cultivation, sites prior to destruction, or hay meadows just prior

Hand-collected seeding spikes of Common Spotted Orchid (*Dactylorhiza fuchsii*) from plants growing in cultivation.

to mowing, all by arrangement and with the permission of the landowner or occupier. You will need plastic buckets or dustbins to collect into. Cut or snap off individual seed heads when the pods have turned from green to yellow, and are yet to turn brown. Place the dustbins or buckets containing the pods in a wind-free spot to dry. A glasshouse or polytunnel is ideal. Turn occasionally, and after two to three weeks the pods will turn brown, will split lengthways and release the minute dust-like seed. When fully dry, crush the pods to release the last of the seed. Clean the seed by passing through a kitchen flour sieve. Take care because the seed will float away in the slightest breeze. Place the seed in a grip-seal polybag. This is best done when the air humidity is low (10–30 per cent relative humidity), as the seed will quickly absorb moisture on a humid day. Store in a fridge.

Machine-harvested Meadow Seed

Surviving unimproved species-rich meadows are a rich potential resource for harvesting meadow seeds. Meadows with a long history of consistent management, spanning perhaps hundreds of years, often consist of communities of species that have evolved as ecotypes with seed maturation dates clustered around the normal cutting date for the meadow. Different long-established meadows will often be either early season, mid-season, or late season heading. Pastures, in contrast to meadows, are normally less species rich. They are also rather indeterminate, with flowering and seed set often spread throughout the growing season. Meadows, compared with most pasture, have a greater potential to produce good yields of very species-diverse seed mixtures. Also, meadows will usually be on ground that can be worked by a tractor, and therefore lend themselves to mechanised seed harvesting, more so than rough pasture. The annual production of wild harvested meadow seeds in Britain and Ireland is around 20,000kg (44,100lb). There is potential for more, but the amount will ultimately be limited by the scarcity of good-quality donor sites.

There are many different ways of harvesting meadow seeds (Gamble et al., 2012). The method employed will depend on the availability of equipment, the site-specific conditions, and the objective. The methods developed and trialled by the author, over many years, are described in the following sections.

Binder

A binder is a side-mounted machine designed to harvest cereals. A reciprocating finger mower cuts at ground level,

Harvesting North Meadow NNR (Cricklade) for seed using a binder in 1982.

Standing sheaves of bunched hay in stooks of three to dry.

the reel (sail) sweeps the cut material back onto the canvas, the canvas delivers this to the buncher, and a knotter ties the bundle and drops the sheaf. The sheaves are stacked by hand in stooks of three in the field to dry. When the sheaves are dry they are carted to the barn to be stored for the winter, or to the yard to be thrown through the threshing machine.

Reaping a meadow is not very different. The vegetation is green so the binder must go more slowly and the sheaves take longer to dry, but the stooks are not ruined by a little rain and seed maturation continues within the sheaves. When dry, the sheaves can be thrashed out by hand on a sheet in the field, or passed through a static combine. The method produces good yields of seed, but is exhausting, and the hay must be disposed of after thrashing.

Green Hay

Green hay is silage, and it is cut, baled, carted and spread by tractor-mounted silage-making machinery. In the meadow, the vegetation is cut by a disc or drum mower (at peak seed maturity), a tedder is used to row up the swath, and the swath is picked up by a baler and tied. Alternatively, a forager can cut, chop and spray the vegetation directly into a silage cart. The green hay is then transported to the restoration site and spread by bale chopper or muck spreader. After drying on the surface of the recipient site and dropping seed, the spent green hay can be grazed *in situ*, or removed and disposed of. If the hay is not removed it will hinder seed germination and establishment because of the physical barrier it presents and because of moderate phyto-toxicity.

The method can successfully transfer seed from donor meadow to receptor site, but the green hay will rapidly heat in transit and must be spread well within two hours of baling, or the seed will heat and die. Also, it is very difficult to quantify the seed content of the green hay and the seed yield can be rather modest. However, green hay can be very successfully used to restore and create meadows over short distances (Wagner et al., 2020; Starr-Keddle, 2022).

Conventional Hay

Conventional hay is preferred to green hay. The donor meadow is cut as normal using a reciprocating finger mower, and the hay is left in the open swathe unturned, to dry. Once dry, the swathes are gently rowed up two into one using an acrobat-type turner going slowly. The row is then gently baled using a conventional small baler or a round baler.

The bales may be directly spread on the receptor site, or may be barn stored until needed. If not spread directly, the seed can be extracted by passing through a static

Mowing hay on North Meadow NNR (Cricklade) with a reciprocating blade finger mower.

Turning hay in the meadow on our home farm. This is believed to have been the first tractor on a farm in our valley.

combine harvester or other cleaner. The seed yield can vary considerably, producing up to 0.4kg (0.9lb) of meadow seed per conventional small bale under good conditions. The drawback of this method is the necessary handling, transport and storage of quantities of baled hay.

Suction Harvesters

Adapted or repurposed leaf and litter collectors have been used to suck up seeds, as have purpose-made seed-suction harvesters. They are all characterised by very low seed-collection rates from meadows and high invertebrate-collection rates. Typically, the hoovered-up seed and accompanying insects eventually pass through a fan and into a collection bag as damaged seed and vaporised invertebrates. Positioning a cyclone before the fan reduces harm to invertebrates (Waring, 1990) and to the seed, but the seed yield still remains very low.

Brush Harvesters

Brush harvesters, as described below, are probably the best option for harvesting meadow seeds. They are tractor mounted, and lift and collect mature meadow seeds by means of a soft, forward-rotating and strategically located and presented brush, without doing harm to invertebrates or damaging the standing crop of hay. They improve on suction harvesters in being kind to invertebrates, easy to move between sites, speedy at harvesting, and high yielding.

The design borrows rear linkage features from side-mounted forage harvesters, harvesting features from horse-drawn Wheat strippers (with rotating wooden pegs), and the Sherborne Reynolds stripper header (with rotating plastic fingers). The first ever meadow brush harvester was the product of the endeavours of the author, and the first year of harvest with this brush harvester was

Home-made suction seed harvester working on North Meadow NNR (Cricklade), 1984.

Brush harvester at Clattinger Farm SSSI.

Suction harvester modified to collect harvested seed in side hoppers by use of an internal cyclone. North Meadow NNR (Cricklade), 1985.

Horse-drawn Wheat strippers from the early 1900s, similar in operation to present-day brush-meadow harvesters.

1986, on North Meadow NNR, Cricklade. Over the decades since 1986 the design of home-produced brush harvesters has changed little.

Critical to performance is the condition of the brushes, their speed of rotation, the forward speed of the tractor, the design of the brush chamber, and the angle of presentation of the vegetation.

Before harvest, the meadow needs to be dry to the touch. It may be wet first thing from overnight rain, morning mist caught in fine leaves and hairs, morning dew formed as moisture from saturated night air condensing on leaves, or guttation from pores in the tip of each blade of grass and leaf tooth (when root pressure below ground exceeds the rate of transpiration through stomata). At harvest time the meadow may not be dry to the touch till midday or later.

Unless brush harvesting is targeting specific species, there is little to be gained from repeated harvesting of the same meadow on successive dates through June and July. For each meadow there is an optimum date on which the greatest quantity and greatest range of species will be obtained. Additional harvests before and after this date add little to the overall yield and range of species (these are personal observations). The timing of spring varies year on year – sometimes it is early, sometimes late – but as summer approaches there is a levelling up and down such that the date for peak seed harvest for classic lowland meadows in England is almost always 11 July. Late June to late July is a busy time for meadow harvesters – and there is no better way to judge harvest date than to field walk.

It is normal to brush harvest clockwise from the edge of the field towards the centre at a forward speed of around 8km (5 miles)/h, depending on terrain. Yield drops off at lower speeds. Avoid unwelcome weeds, especially docks, and avoid lodged vegetation. Unload as necessary onto green sheets (black or clear sheets will burn the vegetation below) in the corner of the field, and allow invertebrates to crawl away or fly off. The brush-harvested material typically consists of 60 per cent seed and 40 per cent chaff by

1.4m (4.6ft) side-mounted meadow brush harvester on the Marlborough Downs.

Loading harvester seed for transport home.

Unloading harvested seed onto sheets to allow invertebrates to escape.

Guttation in Fumitory (*Fumaris officinalis*).

volume, 80 per cent seed and 20 per cent chaff by dry weight. The chaff can be removed in the field by passing it through a pre-cleaner or static combine harvester. The seed may then be dried in the open, subject to weather, and this will take a minimum of three days of good weather, with the seed spread not too thickly and turned regularly.

The seed must be dried down to around 10 per cent moisture content to substantially halt respiration in the seed, in remaining green chaff, and in microbes present in the sample. Seed that is bagged damp will continue to respire, creating and releasing heat and H_20. The temperature will rise in a sack of damp seed, and with each increase of 10°C the respiration rate doubles, until at 40–50°C the seed cooks and dies.

Brush harvesters have a high work rate. The smaller home-produced 1.4m side-mounted brush harvester pictured has a work rate of around 0.8ha/hour (2 acres/hour), and the larger home-produced 3.0m (10ft) trailed brush harvester has a work rate of around 2ha/hour (5 acres/hour). The yield of clean/dry seed will vary hugely between sites, and depending on the time of harvest and the conditions, is typically in the range 25–80kg (55–176lb)/ha (10–35kg (22–77lb)/acre), most frequently around 40kg (88lb)/ha (15kg (33lb)/acre).

Brush harvesting is beneficial to meadow communities because it yields meadow seed that is used to restore and create meadows after decades of harm to them. There is no evidence that the removal of (some) seed from donor meadows changes or harms the donor meadow. Any impact will be small, and is hidden by the more significant impact of the following: the natural waxing and waning of all species, seed predation, pests and diseases, resource depletion, competition dynamics, other environmental and biotic factors, changes in mowing and grazing regimes, and catastrophic events, such as prolonged winter flooding. It is reassuring to part the grass after brush harvesting and count the seeds that have fallen to the ground, normally more than sufficient to ensure annual replenishment of the donor meadow seed bank.

Brush harvesters are now used world wide to harvest wild grasslands for restoration seed mixtures. Some of these harvesters are repurposed leaf and litter collectors, repurposed paddock poo collectors, modified suction harvesters, and adapted bucket brushes.

Pre-cleaning brush-harvested meadow seed by passing it directly through a static combine harvester in the field.

Brush-harvested meadow seed before any cleaning.

3m (10ft) trailed brush harvester.

3m (10ft) trailed brush harvester at work.

Bucket brush modified for harvesting meadow seeds.

Combine Harvester

Conventional and plot combine harvesters are good for harvesting meadow seeds, but the going is slow and a little scary on steep ground. A meadow can be combine harvested directly as a standing crop, but as the vegetation is green it will be slow to pass through the combine and is more likely to block the mechanism than would dry vegetation. The meadow is best swathed first, allowed to dry without turning, and then combined with a draper header fitted. The drum will need to be wide open to reduce shattering of the straw, and aspiration will need to be minimal to retain light seeds and winged seeds, such as those of Yellow Rattle (*Rhinanthus minor*). With low aspiration, movement of material over the frogmouth will be slow and the rate of work will need to be equally slow to prevent clumping of material on the sieve and loss of seed at the rear end. With a 3m (10ft) draper header and a forward speed of, say, 1km/h (0.6mph), the work rate for a combine harvester in a swathed meadow would be, in this example, 0.3ha/hour, just over an acre/hour. This is not far off field experience.

The advantage of using a combine harvester is that it does not require an excess of physical work, and it will produce a clean mixture of seeds. However, it will be rather slow to cover a field unless it is a large machine fitted with a wide header, and it will not work safely on steep ground unless fitted with 'hillside' capability. Of course, donor meadows are sometimes remote, small, and maybe steep or ridge and furrow, all of which are a problem for a combine harvester.

Another method is to use a tractor with a loader and bucket attachment. Drive through the meadow with the bucket positioned low and horizontal, and seed will flip into the bucket. Not a lot, but more if you drive fast.

Crop-grown Native Wild Seed

Native seeds produced in cultivation account for most of the seed used in creating and restoring meadows in the British Isles. There is no central recording of annual production of wild seeds, but an estimate may be gleaned from the acreage known to be under cultivation. From this, annual production of native seeds in the British Isles may be, more or less, as follows:

- 1980: ± 100kg (220lb)
- 1990: ± 1,000kg (2,200lb)
- 2000: ± 10,000kg (22,000lb)
- 2010: ± 50,000kg (110,250lb)
- 2020: ± 100,000kg (220,500lb)

The process of producing native seed in cultivation starts with the collection of stock seed directly from the wild; this is then grown on in a glasshouse or multiplication bed, the multiplied seed is sown to produce field-scale crops, and finally the seed is harvested from these crops, then dried, cleaned and placed in store until required (Knapp and Rice, 1994; Pedrini et al., 2020). For annual species, this may take two to three years, for perennial species the process may take six to twelve years.

Wild Stock Collection

For each species of interest, it is good to identify ecotypes that are well adapted for the intended restoration use, are genetically robust, and are representative of the wild type of the species. The next step is to identify the location of a large and representative wild population of that ecotype. A population is a freely and actively interbreeding cluster of individuals, and walking the site the limit of the outer boundary of the population can usually be assessed. Stock collection may reasonably aim to capture 95 per cent of the genetic variation present in the population. This may be achieved by walking over the site of the population in a random, spaced, zigzag manner, collecting seed by hand from at least fifty individuals, while minimising conscious and unconscious selection (Marshall and Brown, 1975). It is good to time collection to coincide with peak seed maturation, thus reducing selection for early and late flowering genetics. Also, complete a recording sheet while on site, capturing the data that fully describes the sample, supported by images (see table on the right).

Take care to restrict sampling to no more than 20 per cent of the seed produced by the population, as a safeguarding precaution. However, the greater the quantity of wild stock seed collected, the greater the quantity of genetic variation captured. Take, for example, Cowslip (*Primula veris*), which has a seed count of around 900 seeds/gram. A sample size of 0.1g (0.0035oz) risks inbreeding depression in the subsequent propagation line, while 10g (0.35oz) does not. The quantity of the initial sample also influences the number of generations required, and the time taken, to reach full field-scale seed production. Cowslip has a generation time of ± three years in cultivation (longer in the wild). The larger the initial

Hand collecting stock seed.

FIELD RECORDING SHEET

SPECIES LATIN		COLLECTOR	
COMMON NAME		EMAIL	
		PHONE	
ACCESSION No.		ADDRESS	
SITE LOCATION		POPULATION DESCRIPTION	
County			
Nearest town		Age	
Nearest village		Number plants	
Place name		Area	
Field name		Density	
Status (SSSI etc)		Flower type	
Grid ref		Leaf type	
Landowner		Seasonality	
		Other	
SITE DESCRIPTION			
Habitat		COMMUNITY DESCRIPTION	
Soil type			
Bedrock type		Vegetation type	
pH		Community type	
Fertility		Community age	
Wetness		Community size	
Altitude		Naturalness	
Slope		Management	
Aspect		Associated sp	
SAMPLE DATA		OBSERVATIONS	
Date collected			
Seed productivity			
Seed maturity			
Collection method			
No. plants sampled			
Date seed cleaned			
Quantity obtained			
Date in store			

seed sample, the fewer generations required to reach full-scale field production. For Cowslip:

- 2g (0.07oz) sample = three generations (nine years) to first field-scale harvest of seed
- 20g (0.7oz) sample = two generations (six years) to first field-scale harvest of seed
- 200g (7oz) sample = one generation (three years) to first field-scale harvest of seed.

2g (0.07oz) of stock seed may take one hour to collect, and 200g (7oz) of stock seed may take two hours to collect. The extra time taken to collect 200g saves six years on the production cycle of Cowslip. This principle has wide application, and for most species it is good to aim to collect sufficient stock seed to enable direct establishment of field-scale seed production, without the need for intervening multiplication.

Ownership of the stock seed is usually acquired upon collection, but as the seed is going to be used for propagation purposes, and may be sold, ownership of the genetics within is a moot point. Breeders' seed contains intellectual property (IP) that is owned by the holder of the plant variety rights (PVR) for the cultivar. Someone can own the seed of a cultivar, but ownership of the genetic IP within remains with the PVR holder. However, it is different for wild native species. The wild genetic resource of these islands is ultimately the property of the people of these islands, and at present it is free for us to multiply and use without restriction, unless prevented from doing so by Schedule 8 of the Countryside and Wildlife Act 1981. This is good. Nevertheless, it is understandable that countries within the British Isles may wish to maintain the separateness of their national or regional flora, and we should respect this. More broadly, the Convention on Biological Diversity, Rio de Janeiro (1993) commits 190 signatory countries to 'the sustainable use of (biodiversity), and the fair and equitable sharing of the benefits from the use of genetic resources'.

Multiplication

If the quantity of stock seed collected is small it may need to go through one or more generations of multiplication and bulking up. For very small quantities of stock seed the only option is to raise plants using conventional horticultural methods. Sow the seed in the autumn on the surface of damp, peat-free compost in a seed tray, sprinkle with grit, cover with a pane of glass and place in a cold frame or outside against a north-facing wall. At the first signs of germination move the trays to a cold glasshouse, remove the pane of glass and prick out with a pencil into trays or pots, holding each seedling gently by a cotyledon. Harden off in the spring and plant out in rows on prepared ground, and grow on until harvest. If necessary, repeat, direct drilling the second and/or third generation using seed harvested from the previous generation, and by this means bulk up the quantity of propagated seed. Quantities of stock seed of 10g (0.35oz) or more can be direct drilled with a precision drill into prepared ground, and this method of establishment (for bulking up) is preferred over plant raising and planting out.

Cropping

Seed-production crops of native restoration species may be established by planting out or by direct drilling, and the crop may be grown on and maintained 'weed' free by precision cultivation or herbicides, or neither of these but grown with a nurse cover. Different aspects of these methods may be selected and merged depending on species, farm capability and operator preference. Ultimately, large-scale, nature-friendly cropping will most often consist of direct drilling into prepared ground with a managed annual nurse, and without the use of chemicals.

Planting out: Raise plants so that each planting unit is derived from one seed. Plant out into prepared soil in rows at a standard row width equal to one third of the wheel spacing of the farm machinery, or select a different width and adjust the wheel spacing of the farm machinery to take three rows between wheels. The between plant spacing within rows will differ from species to species according to mature plant size, and should be sufficient to minimise competition between individuals, while making best use of the ground.

Direct drilling: Prepare the ground by cultivations and rolling, avoiding ground that carries a legacy of residual herbicide from previous cropping and/or troublesome perennial weeds. Aim to produce a medium/fine firm tilth, using a ribbed roll for the final roll, rolling at an angle to the row direction. Select a time for drilling that is consistent with the seed dormancy and life-cycle features of the sown species. Precision drill in a standard row width (for example 58cm (23in)), with between-plant spacing determined by plant size at maturity. Drill across the contours (on sloping ground), with coulters set to place the seed on the surface or at up to 10mm (0.4in) deep according to the species, and drop five to eight seeds at each within row location. For very fine seed, dilute with cooked commercial Brown Top (*Agrostis capillaris*) seed. Finally, roll with a Cambridge (ribbed) roll to firm in the seed.

Precision cultivation: Maintain crops 'weed' free by between-row cultivation by steerage hoe or rotary cultivator, and between plants/within rows by hand hoeing. Alternatively, precision drill using a robot and let the robotic weeder keep the crop clean.

Herbicides: At a conference back in the 1980s, following a presentation by Dr Derek Wells of the Nature Conservancy Council, a member of the audience asked if herbicides could be beneficially applied to the cropping of restoration species for seed. Derek replied that there is no place for herbicides in nature restoration as herbicides are designed to kill the very plants we are trying to save. The same argument may well be applied to the use of insecticides, fungicides, nematocides and molluscicides. Current rules restrict the application of pesticides and selective herbicides to only those crop species that are specified on the label. Most native species are not listed as permitted crops for application, and therefore pesticides and selective herbicides may not be used on most native species in cultivation. Off label use of pesticides and selective herbicides is an offence. Non-selective herbicides are permitted, and may be used at the discretion of the grower, so long as the label instructions are followed.

Production rows of Red Fescue (*Festuca rubra*) established directly from wild collected stock. The annual weed cover acts as a beneficial nurse.

Nurse cover: The 'nurse' method relies upon ground cover of nurse species arising naturally from the soil seed bank, or from seed broadcast just prior to drilling/planting of the perennial crop. The annual nurse, if it is a desirable restoration species such as Corn Poppy (*Papaver rhoeas*), may be harvested for seed before the perennial row crop reaches maturity. No attempt should be made to maintain bare soil within the crop. Cleanliness of the harvested crop is achieved by investment in precision seed-cleaning technology, rather than in pesticides and chemicals. And experience tells us that the presence of nurse species reduces crop susceptibility to pests and diseases.

Harvest

The method of harvest will depend on the species. Some species may be best harvested by hand, either because the seed capsules are poorly presented (*Primula vulgaris*), because seed maturity is quickly followed by dehiscence (*Geranium pratense*), or because seed maturity is indeterminate (*Knautia arvensis*).

Combine harvesting is the preferred method for most wild species. Cereals have been bred to retain ripe grain attached to the inflorescence (to increase harvestable yield), but most native species do not retain seed for long after maturity. Accepting that a considerable amount of early and

Established row crop of Cowslip (*Primula veris*) in flower.

Hand harvesting seed of Field Scabious (*Knautia arvensis*).

late seed will not be harvested, field walking and looking at the crop and the weather are the best way to judge when to go in with the combine. Each crop will need a different combine setting, some crops will benefit from swathing, some may need picking up twice, and all will be slow to combine and many will repeatedly block the drum and walkers.

Harvested seed will need to be laid out and dried before bagging. Drying will take a minimum of three days in the open, taking longer on cool, still days with high humidity.

Combine harvesting a crop of Sweet Vernal Grass (*Anthoxanthum odoratum*).

Combine-harvested seed set out to dry.

Small quantities of hand-picked seed drying in open buckets, seen with the kitchen sieves used for pre-cleaning small amounts of seed.

For perennial restoration species the maiden harvest, in year two or three, may be modest, followed by the main and usually highest-yielding harvest, this then followed by subsequent harvests for several years that may tail off due to the accumulation of pests and diseases, or may not.

Seed yields of native species in cultivation are normally in the range 10–100kg (22–220lb)/acre.

Genetics

The wild stock of a species may normally remain in cultivation for up to five generations and still retain 95 per cent of the genetic variation present in the original wild collection (Marshall and Brown, 1975). Huge field-scale populations, and spacing between individuals to reduce competition, will slow the rate of loss of genetic variation (Lauterbach et al., 2012). Further, this loss may be more than compensated for by introgression from nearby wild populations of the same species within pollinator range.

More noticeable will be the inevitable shift in population mean during cultivation in response to natural selection and unintended artificial selection, during propagation. Flowering date, seed maturation date, seed size, seed germinability and ease of establishment are all highly responsive to selection pressure under cultivation. The greater the number of generations in propagation, the greater the shift in the population mean (MacIntyre and Elliott, 1974; Conrady et al., 2023; Nagel et al., 2018). For this reason it is good to replenish stock after around five generations of propagation. However, it is good to remember that shifts in mean during cultivation can bounce back in nature so long as the genetic variance remains high.

CHAPTER 12

SEED PROCESSING AND SEED CARE

In this chapter we look at methods for drying and cleaning harvested seeds, and the best storage conditions for maintaining viability. We also discuss seed dormancy and seed germination.

Seed Processing

The vast majority of British and Irish restoration species have orthodox seeds that may be dried without harm. Recalcitrant seeds, those that will not tolerate drying, are more typical of tree species, although a few herbaceous species have recalcitrant seeds – for example, Wood Anemone (*Anemone nemorosa*). Harvested seed of orthodox species must first be dried to substantially halt respiration within the seed sample, allowing it to be stored without heating.

Seed cleaning allows the removal of weed seeds, chaff and empty seeds. For small lots, kitchen sieves or mini versions of aspirated vibrating sieve machines are ideal. To handle larger lots, a seed processing plant may be usefully equipped with a range of machines including the following:

- De-awners and de-huskers
- Aspirators and aspirated vibrating sieve sorters
- Indent cylinders and gravity tables
- Colour sorters and robotic sorters

The skilled operation of these machines allows mixtures of species to be separated into their component species and allows the removal of chaff, weeds and non-viable small or empty seeds. A minimum standard for cleaned seed is 95 per cent purity and 90 per cent viability, although in practice a much higher standard is reasonably easy to achieve.

Seed-cleaning machines.

Seed Longevity and Storage

A special feature of seeds is that they are able to suspend physiological processes and survive dormant for very long periods without loss of life. Our familiar Corn Poppy (*Papaver rhoeas*) has tiny seeds that are extremely long lived when buried in the soil. Their half-life is forty years, so even if buried for centuries below grassland, when the ground is disturbed some seeds may be brought to the surface, and when exposed to light, they germinate, grow and burst into flower.

Longevity of seed varies between species; Corn Poppy has very long-lived seed and Yellow Rattle has very short-lived seed. The moisture content of the seed and the temperature at which the seed is stored also influences longevity. The moisture content responses and the temperature responses vary between species, and moisture content and temperature minimum tolerance interact one with another. As a general rule, orthodox seeds have very good longevity if stored at around 15–20 per cent relative humidity (10–15 per cent moisture content) and at 1–3°C (Vitis et al., 2020). A dehumidified cold store at these settings provides optimal conditions for seed stored in breathable sacks. For seed stored in sealed containers it can be pre-dried to equilibrium at 15–20 per cent relative humidity before sealing in glass vials and placing in a cold store. Ultra-low temperature and ultra-low moisture content can improve longevity even further for some species, but the tolerance range and longevity response to different ultra-low moisture contents and temperatures varies from species to species.

Eventually, when a seed lot deteriorates, it does so slowly. At above optimal moisture content for storage, oxidative processes damage DNA, physiological repair mechanisms do not function properly, and spontaneous mutations and harmful chromosomal aberrations accumulate. Lettuce (*Lactuca sativa*) seed transferred from sub-optimal damp cold storage to fully imbibed cold storage improved in germinability over time and chromosomal aberrations declined, suggestive of operative repair mechanisms. Surprisingly, Lettuce seed stored fully imbibed/cold has better longevity than seed stored under optimal dry/cold conditions (Villiers, 1974). It may be deduced from this that longevity of seeds in the soil seed bank is best in soils that are very dry most of the time, and also in soils that are very wet most of the time. The seed bank in soils that are intermediate in wetness may, possibly, have the poorest seed longevity and may carry a higher load of spontaneous mutations than otherwise. Of course, soil moisture contents fluctuate, and the range and frequency of fluctuations may also impact on the longevity of the soil seed bank.

The frequency of chlorotic (non-green) seedlings arising from the germination of a seed lot is a measure of the frequency of spontaneous mutations in a seed lot, and this frequency increases as viability of a seed lot declines. Speciation in apomictic species is driven by rates of occurrence of spontaneous somatic mutations in populations, and also by the fixation rate (in the population) of any of these spontaneous mutations that may be advantageous. It is interesting to speculate that the geographic location of speciation hot spots in apomictic species, such as Hawkweeds (*Hieracium sp*), may correlate with locations having intermediate and fluctuating soil moisture contents and warm soil temperatures.

Seed Dormancy and Viability

Seed dormancy mechanisms have evolved to ensure that the state of readiness to germinate is spread over time (not all seeds being ready to germinate at once), and that peak germination readiness coincides with favourable conditions and at a favourable time of year (often spring or autumn) (Kildisheva et al., 2020; Smith and Weller, 2020; Fenner and Thompson, 2005). Vernalisation, the requirement for seed to be chilled while fully imbibed, is a feature of the seed of Cowslip (*Primula veris*) and Yellow Rattle (*Rhinanthus minor*). Hard-coated (hard testa) seeds require a period of abrasion or exposure to microorganisms in the soil to degrade the hard seed coat, and this is a feature of Meadow Crane's-bill (*Geranium pratense*) and Tufted Vetch (*Vicia cracca*). Some species have no seed dormancy and the seed is ready to germinate at maturity. This is a feature of Moon Daisy (*Leucanthemum vulgare*) and Common Knapweed (*Centaurea nigra*).

Dormancy requirements have been bred out of the cultivars of most crop species, but it is an important feature of wild species, helping to enable the success of restoration seeding. Agricultural seed enhancement treatments such as seed priming, seed scarification, seed pre-chilling, seed pelleting, and chemical seed coatings are almost certainly of no benefit to the restoration process and may hinder natural processes (Pedrini et al., 2020, 2021; Butler et al., 2009).

Seed fill may be assessed by a cut test with a razor, or by X-ray imaging, to judge the percentage of filled seeds, and a tetrazolium test will give an indication of the presence of living tissue. The ultimate viability test is germinability.

Seed Germination

For almost all native wild species there is a set of common requirements, all of which must be met simultaneously, to trigger seed germination:

- The seed must be non-dormant
- It must be viable (alive)
- It must be fully imbibed
- The ambient temperature must be in the range required by the species
- The seed must be exposed to light (a requirement bred out of most crop cultivars).

Germinability of a sample may be tested by placing layers of wet kitchen roll in a seed tray, sprinkling the seed sample thinly on the surface (so that seeds are well spaced from each other), covering with clingfilm, and placing on a window sill. Count and remove seeds as they germinate until no more germinate. The remaining seeds will be dormant or dead. Dead seeds will often develop a halo of fungal growth.

High standards for the production, recording, processing, storage and testing of native seeds are essential for restoration success (Oldfield et al., 2019; Frischie et al., 2020).

CHAPTER 13

GROUND PREPARATION AND SEED SOWING

In this chapter we look in detail at the practical methods that may be used to prepare ground and sow seeds; we also discuss the all-important topic of seed rates.

Ground Preparation for Restoration Methods

Restoration methods are those methods used to diversify existing vegetation by adding seeds without the need to destroy the existing vegetation. The ground preparation is the same for meadows, pasture, and most other types of herbaceous wild vegetation. Cut and remove the existing vegetation in late summer/early autumn. At a field scale this could be a late hay or silage cut. On smaller plots mow or scythe and rake off. If the ground is grazed, graze down as much as possible. Then scarify to remove the dead thatch on the soil surface. Thatch can be phytotoxic and can reduce germination and establishment of the sown seed (McKenzie-Gopsill and Farooque, 2023; Mazzoleni et al., 2015). Continue scarification, aiming to create up to 50 per cent bare soil within the vegetation. Scarification options include the following:

- Raking: Good for hand scarification of small areas.
- Grass harrows: Both spring tine and fixed tine harrows are good for scarification of large areas. They will not harm the existing vegetation, they remove thatch and they create sites for seed germination.
- Disc harrows: Good for scarification of large areas. Will check the existing vegetation by breaking some root connection with the soil, and will create sites for seed germination. Beware deep discing on very dry ground as this may check the existing vegetation more than needed and leave a lumpy surface.
- Power harrows: Not good for scarification because power harrowing will over-harm the existing vegetation. The process of power harrowing is tantamount to ploughing, and as such may fall foul of the EIA regulations, which are designed to protect permanent semi-improved and unimproved grasslands in blocks greater than 2ha (5 acres).

Ground Preparation for Creation Methods

Meadow creation methods are methods used for creating wild herbaceous vegetation by sowing a complete mixture of native species onto bare prepared soil. This may be bare ground created by engineering work, and if it is, it is best not to finish with a layer of topsoil. It may be former arable land, and if it is coming out of cereals beware any carryover of residual herbicides. Land with vegetation that has no retention value can be cultivated to remove the existing vegetation. This may require repeat cultivation. The conditions of the site will determine the type of cultivations used: it may be plough, tine harrow, disc harrow or power harrow, or a combination of these. The aim is to produce a weed-free, firm seedbed of medium to fine soil particles.

Seed Rates

The standard sowing rate for a perennial meadow mixture in a garden, park or landscaped setting is 4g/sq m (1.4oz/10.7sq ft). This applies for sowing both into existing scarified vegetation (restoration) and for sowing onto bare soil (creation). This rate is sufficient to get the best and quickest results on ground that may sometimes be less than ideal. It is important not to exceed 4g/sq m because this can lead to over-dominance of the more quickly establishing species in the mixture. Annual mixtures sown in a garden, park or landscaped setting may be sown at a standard rate of 2g/sq m (0.7oz/10.7sq ft) (Lang et al., 2016).

On farmland and in the wider countryside the standard rate for a perennial meadow mixture is usually 2g/sq m (0.7oz/10.7sq ft). Ground conditions on farmland are usually good, and 2g/sq m is sufficient to produce good results. If the meadow seed available is limited in quantity, and the area to be sown is large, it is better to sow a larger area at a further reduced rate (say, 1g/sq m (0.4oz/10.7sq ft)) than it is to sow half the area at 2g/sq m. The difference in outcome is that acceptable ground cover may be achieved after two years at 2g/sq m and after three years at 1g/sq m, the lower rate relying more on the establishment of second-generation plants within the community to achieve a satisfactory outcome.

Ultra-low seed rates of, say, 0.1g/sq m (0.04oz/10.7sq ft) and 0.01g/sq m (0.004oz/10.7sq ft) can also be successful (with favourable management in favourable conditions) in establishing species-rich meadows and other herbaceous communities, but the time taken will be longer (ten to twenty years) as there will be greater reliance on second, third and fourth generation build-up of Yellow Rattle (*Rhinanthus minor*) and the perennial native species within the community.

When to Sow

If the success of the project is reliant on the establishment of Yellow Rattle, the meadow mixture must be sown in the autumn. All other mixtures can be sown at any time of the year when ground conditions allow. The time to first flowering subsequent to autumn or spring sowing in the open will vary from species to species depending on vernalisation requirements (to break seed dormancy), seasonality of germination, typical time from establishment to first flowering (annual, biennial, perennial), and, of course, site and weather conditions. Averaging over all perennial species in a meadow mixture, you may expect germination times and rates as described below.

Autumn Sowing

Sowing in September usually allows some pre-winter germination, and these plants can flower weakly in the following summer (year two), but the main maiden and first flowering of the perennial community will be in the next summer (year three). First flowering of the community is dominated by fast establishing species, with species that establish more slowly flowering in later years.

Sowing in November will not usually allow pre-winter germination. There may be expected to be no flowering of the perennial community in year two, with full maiden flowering in year three. A late autumn sowing may be expected to produce better spring germination, and greater species diversity, than a spring sowing.

Spring Sowing

A March sowing will produce rapid ground cover but no flowering of the perennial community or Yellow Rattle in year one. Full maiden flowering may be expected in year two.

How to Sow and Rolling in

The prepared soil should be friable and of a medium to fine tilth. The seed is broadcast on the surface and falls between the soil particles. If drilled in below the surface, or

Meadow seeds broadcast onto scarified grassland using a tractor-mounted fertiliser spinner.

3m (10ft) roller feed drill designed for meadow seeds and fitted onto 3m discs with packer wheel, placing the seed on the surface of the prepared soil between the discs and the packer wheel.

incorporated by harrowing, the seed will be buried too deep to receive the light required to trigger germination. Broadcast the seed by hand from a shallow bucket using a seed fiddle or a seed barrow. A tractor-mounted spinner can be used, but it is difficult to control the rate applied and the evenness of spread. Some agricultural/amenity seed drills will handle mixtures of seeds at low rates, but some cannot. The 3m (10ft) roller feed drill pictured can sow a meadow mixture evenly at down to 0.1g/sq m (0.04oz/10.7sq ft). Roll after sowing to give good seed/soil contact. It is a good idea to retain the bag label and a sample of the seed, and also to mark out a small control plot that is not sown, for comparison later.

3m (10ft) roller feed drill sowing onto scarified grassland in one pass.

Precautions against seed predation are not normally necessary because the seeds are very small and the normal sowing rate too low to be attractive to seed-eating birds. Nor is watering required because germination will be spread over the twelve months following drilling, and germination will naturally occur when the soil moisture content is sufficient.

First Signs of Germination

From an autumn sowing of a complete meadow mixture sown onto prepared soil (meadow creation) the first species to germinate will be annual 'weeds' present in the soil seed bank plus any annual nurse species that have been sown with the meadow mixture. It is important to leave this growth uncut and ungrazed until the following

Cambridge ribbed rolls for firming in meadow seeds after sowing.

Germinating seeds, some of which will be annuals from the soil seedbank.

Increasing ground cover, provided by annual species at first.

Established ground cover from the sown meadow seeds.

mid-summer, so that it can act as a nurse to the slower germinating and establishing perennial meadow species. The same applies to a spring sowing: leave uncut and ungrazed until mid-summer of the same year.

Germination of the perennial meadow species will occur throughout the seasons of the first twenty-four months after sowing, some species earlier than others, and most in the first twelve months. Parting the annuals and taking a close look at the ground will reveal the meadow seedlings, but at this stage it takes an expert eye to identify the species. The perennial seedlings will mark time nursed by the annuals. When the annuals die back or are cut back in summer the germinated meadow species, then exposed to sunlight, put on a spurt of vegetative growth before winter, building capacity to flower the following summer.

From an autumn sowing of a complete meadow mixture into scarified existing species-poor amenity or agricultural grassland (meadow restoration), the first species to germinate will be the annual, Yellow Rattle, in the following spring. The seedlings of Rattle have distinctly puckered first true leaves set between a pair of smooth cotyledons. Germination of the sown perennial meadow species will occur throughout the seasons of the first twenty-four months after sowing, some species earlier than others. It is not normally possible to detect germination of these species as seedlings amongst the existing grass, but they will become apparent later as they grow and flower.

CHAPTER 14

MANAGEMENT METHODS AND TOOLS

In this chapter we look at general methods for managing meadows and other herbaceous vegetation. The intensity and mix of management methods required will vary between habitats and will be tempered by site and machinery constraints. We look in detail at mowing, grazing and weed control.

Mowing and Cutting

Silage making has mostly replaced hay making on commercial livestock farms in Britain and Ireland. The few hay meadows that remain are usually cut in June/July, the hay or haylage is dried in the field, baled/wrapped and carted to the barn to be stored to provide winter feed for the farm livestock. Maintenance of high species diversity in these established meadows is achieved and sustained by the annual cut and removal of the hay/haylage in summer, followed by aftermath grazing. Survival of a meadow depends on this hay-making cycle and the species composition of the meadow depends, above all, on the cutting date. It makes no significant difference if the forage is made into hay, haylage or silage.

Cutting Date

The farmer will know if his/her species-rich old meadow is early, mid-season, or late heading (flowering). Consistent management over generations interacting with local conditions will create these differences. The gate to North Meadow NNR, Cricklade, is unlocked on 1 July for mowing. The meadow has a recorded history going back 800 years, and it is not surprising that the vegetation has evolved to be ready for mowing in the first two weeks of July, in most years. This would be considered to be mid-season for a lowland unimproved meadow in England. Knowing the normal heading date for the meadow, the farmer can time mowing for peak yield, or can go in earlier for a cut that is low in fibre, or later for a cut that is high in fibre, depending on the stock being fed.

This aside, the cutting date may also be varied for reasons related to the benefit to nature:

- A mid-June cut favours early seeding species such as Dandelion (*Taraxacum officinale*), Bulbous Buttercup (*Ranunculus bulbosus*) and Sweet Vernal-grass (*Anthoxanthum odoratum*), but is not good for most other meadow species, ground-nesting birds and invertebrates.

It is, however, a good date to cut to suppress Docks (*Rumex sp*), Thistles (*Cirsium sp*) and Nettles (*Urtica dioica*), and prevent return of their seed to the ground.

- A Summer Solstice cut (midsummer's day) removes the greatest amount of nutrients from the ground with the hay; less is removed before or after this date, but the date is too early for some meadow species, ground-nesting birds and invertebrates.
- An early July cut is good for most early and mid-season meadow species such as Yellow Rattle (*Rhinanthus minor*), Moon Daisy (*Leucanthemum vulgare*) and Knapweed (*Centaurea sp*). There are less likely to be ground-nesting birds at this time, but invertebrates will still be abundant and active.
- 15 July is the stewardship date after which meadows in agreement may be cut. This date is good for wildlife, although a little late for the farmer, but on balance it is a cutting date that has proven to be of the greatest all-round benefit.
- A late July cut will favour later meadow species such as Meadowsweet (*Filipendula ulmaria*), Devil's-bit Scabious (*Succisa pratensis*) and Great Burnet (*Sanguisorba officinalis*).
- An early August cut will allow grassland Orchid species to shed seed, and will favour very late species such as Saw-wort (*Serratula tinctoria*), Lady's Bedstraw (*Galium verum*) and Wild Marjoram (*Origanum vulgare*). By this time the hay will be very fibrous unless there has been some good bottom growth.

Once the meadow is cut, the shelter and food source for pollinators, other invertebrates, birds and mammals is gone. They perish or seek out a new home. This is unfortunate, but cutting is necessary to ensure the continued well-being of the meadow. However, a compromise is possible. Leave some corner or edge of the meadow uncut as a refuge for the autumn and as a hibernation spot for the winter. If the refuge is rotated it will not have much of an impact on meadow diversity, but if it is at a fixed location each year it will soon become gloriously tussocky and wild.

The above deals with cutting dates for meadows in an agricultural setting. In the garden, a landscaped setting, a park or on a roadside, cutting a meadow is a little different. 'No mow May' has transformed many domestic lawns into flowering lawns, and some gardeners may now be tempted to 'No mow April, May and June', and that is good. Whether additional species are added by patch scarification and seeding or not, the lawn thus transformed into a meadow is a delight. Eventually it needs to be cut and the hay removed, but the date of that cut will be very much a personal matter.

Meadows in parks, landscaped settings and on roadsides are best cut annually to maintain and improve species diversity. A late June cutting date for the meadow will remove vegetation before it starts looking scraggy, but this is too early for invertebrates and ground-nesting birds. A middle of July cutting date is often the best all round. The cut can be left until August or later, but the meadow will look increasingly unkempt and rather rank.

Rankness is an over-dominance of coarse grasses such as Cocksfoot (*Dactylis glomerata*), Yorkshire Fog (*Holcus lanatus*), False Oat Grass (*Arrhenatherum elatius*) and other excessively robust species. It is often a consequence of late cutting and may be partially rectified by cutting in June for a year or two, then reverting to a July cut.

Meadows sown on infertile mineral soils, such as pure chalk, sand or limestone subsoil, are a special case. They

Mowing into swathes with a tractor in preparation for hay making.

Mowing by strimming. On this verge the hay is stacked ready for carting away loose.

will take years to establish and will probably never produce sufficient vegetation to merit cutting every year. These meadows can be cut on a five- to ten-year cycle, the main purpose of which will be to control scrub.

Cutting Methods

Meadows in an agricultural setting may be cut with a tractor mower, either an old-fashioned reciprocating finger mower or a modern drum/disc mower. Mowing produces swathes (lines of cut vegetation) ready for turning, rowing up and baling into hay. Swathes can also be cut by reverse driving a twin rotor topper. Alternatively, a meadow can be cut using a self-propelled or tractor-mounted forager, delivering the cut green hay directly into a trailer. A flail mulcher or hedge trimmer may be used, but only in situations where the vegetation is so sparse that it does not merit picking up after cutting, and provided that the mulched vegetation, when left on the ground, is not sufficient to kill the vegetation below.

In some years, because of inclement weather or for other reasons, it may not be possible to cut the meadow at any time. An option here is to graze the standing hay to thin it out and then to top the residue. On a thin hay crop this can be as beneficial to the meadow as a traditional hay cut. On a moderate to thick hay crop, grazing and topping is a second best to the traditional hay cut, but better for the meadow than leaving it uncut into the autumn (FMP newsletter, 2013).

In the garden a scythe is handy. Keep the blade sharp, and cut low. In the hands of a skilled scyther, a scythe is equal to a petrol strimmer, and greener and more satisfying to use.

Scything can also be a way to cut a meadow in a park or in a landscaped setting. Alternatively use an Allen scythe, or a self-propelled 'cut and collect' machine, which will do the work in one pass. These machines come in a variety of models and sizes.

Mowing by hand scythe. The hay is removed green for composting or for feeding livestock.

Disposal of Cut Vegetation

The cut hay must be removed from the meadow, and if it is not going to be used as feed, it is best composted at a permanent spot for composting nearby, but well away from a river, ditch or pond. Stack the hay, dry or green, and let it slowly rot over several years, creating a carbon store to offset against other activities, and creating a welcoming habitat for microbes and creatures great and small. Do not dry and burn it, as this will kill all life in the hay and unnecessarily release CO_2 to the atmosphere along with toxic chemicals and particulates.

Grazing

If the created/restored native vegetation is grazed, the choice of livestock and their management can have a huge impact on the diversity and composition of the vegetation.

The milking herd on the home farm, sometime in days gone by.

Cattle

Cattle are ruminants with no top teeth up front. They graze by grasping the greenery with their tongue and clasping it between their top pad and lower front teeth, and pulling. They are not selective grazers, but they will avoid Ragwort (*Jacobea vulgaris*), Nettles (*Urtica dioica*) and Spear Thistle (*Cirsium vulgare*), otherwise eating everything that is in front of them. The effect they may have on the composition of wild vegetation is probably fairly neutral. They are

Extensive grazing with rare breed cattle.

very efficient at removing forage, and being ruminants they are very efficient at extracting nourishment (by fermentation) from the forage they eat. Native breed beef cattle can make a positive contribution to the conservation management of aftermath grazing and extensively grazed pasture. They are particularly useful on wet ground that may not be good for sheep and some equines.

Sheep

Sheep are ruminants with only bottom teeth up front. They graze by nibbling vegetation and are highly selective and efficient grazers. They prefer flowers to foliage and grass, and will strip a summer meadow of flowers in a blink. They next turn to the foliage of wild flowers, nibbling down the stem and then nibbling out the central growing point of rosette species such as Cowslip (*Primula veris*), Rough Hawkbit (*Leontodon hispidus*) and even Ragwort (*Jacobea vulgaris*). They have little time for grass. Sheep are nothing short of a disaster for biodiversity, having stripped areas of the uplands and lowlands of anything that is not grass. They do not poach the ground, but create a tight grass-dominated sward with little opportunity for species recruitment. Summer sheep grazing is the most harmful to biodiversity, winter sheep grazing is rather less so. If sheep are the only grazers available, it is better for the well-being of a meadow to graze the winter aftermath with sheep than to leave it ungrazed. The easiest gain for biodiversity in Britain and Ireland would be to reduce sheep-grazing pressure in the uplands.

On the farm, the feed value of cattle pasture can benefit from winter sheep grazing. The sheep eat the wild flowers and weeds in winter, consequently increasing the grass content of the spring sward, when the cattle are turned out. There is a long tradition of bringing sheep from the uplands to the lowlands for winter grazing, then returning them to the uplands in the spring.

Ewes and lambs grazing in spring.

Equines

Equines have top and bottom teeth up front and an exceptionally sensitive and large muzzle. Like cattle, equines cannot see what they eat because their nose gets in the way (Hanggi and Ingersoll, 2012), but unlike cattle, the equine muzzle is covered in fine sensory bristles and hairs allowing the vegetation to be scanned, informing the choice of species that are consumed. Generally, equines have a grazing preference for grass species, and will pick out grass from among the flowers. Year on year managed equine grazing of aftermath and extensive pasture will generate species-rich swards with a low to moderate grass content. They do not graze as close as sheep and do not nibble out the growing point of rosette species. Moderate winter poaching with equines may be beneficial for meadows as it can promote recruitment of native plants from the soil seed bank.

Equines are non-ruminants, and compared to cattle and sheep, require double the intake of forage to gain an equal amount of nourishment, and are minor emitters of methane. They only partially digest their forage and return large quantities of high-fibre biomass to the ground via biodeposition. Cattle convert most of their forage intake to milk, weight gain, CO_2 (from respiration) and CH_4 (methane from digestion) and are net releasers of greenhouse gases to the atmosphere. Equines help remove CO_2 from the atmosphere by the role they play in the carbon cycle, contributing hugely through biodeposition to the carbon stored in biomass in the soil.

Aftermath grazing does not provide sufficient time for equine latrines to form, and latrines are not a feature of meadows, but they are a feature of pasture grazed year-round by equines. Females and geldings check out

Extensive conservation grazing with Shire horses on the home farm.

Foals mutual grooming.

EXAMPLE SPECIES SELECTIVELY GRAZED BY EQUINES

- Most grass species
- Cow Parsley (*Anthriscus sylvestris*) and Hogweed (*Heracleum sphondylium*)
- Dandelion (*Taraxacum officinale*)
- Cleavers (*Galium aparine*) and Traveller's Joy (*Clematis vitalba*)

EXAMPLE SPECIES SELECTIVELY AVOIDED BY EQUINES

- Docks (*Rumex sp*)
- Thistles (*Cirsium sp*), but will nibble individual flowers in bud
- Stinging Nettles (*Urtica dioica*), but will eat late-season senescing Nettles
- Ragwort (*Jacobea vulgaris*), and will stamp on Ragwort in disgust
- Clovers (*Trifolium sp*) and Buttercups (*Ranunculus sp*)
- Moon Daisy (*Leucanthemum vulgare*) and Knapweeds (*Centaurea sp*)
- Meadow Cranesbill (*Geranium pratense*) and Wild Marjoram (*Origanum vulgare*)
- Yellow Rattle (*Rhinanthus minor*) and Betony (*Betonica officinalis*)

Equine droppings showing partial digestion of forage and germination of viable seeds.

droppings left on the ground by others, then add their own without stepping forwards, and so a latrine patch develops. Entire males exhibit different behaviour. They check out droppings at approach points to their territory, step forwards, then add their own, and so a pile develops, the point of which is to be on top. Latrine patches and piles contribute structural diversity to pasture, are a bio-deposition carbon store, and support a host of invertebrates including dung beetles.

In addition to all of the above, equines moderately poach the sward, creating opportunities for species recruitment. Also, they disperse ingested seeds (Cosyns and Hoffman, 2005) and native breeds are hardy and require minimal management on open ground. They are also exempt from livestock movement tracking documentation and are resistant to bovine TB, mad cow disease (BSE) and foot-and-mouth disease (FMD). They seem ideally suited for conservation grazing (Gewin, 2020). It is only for reasons of respect and tradition that they are not eaten in the UK.

Weed Control

Annual weeds are beneficial as a temporary nurse to newly sown perennial meadows, and naturally decline and give way as a perennial meadow establishes. Annual weeds are not a significant problem in established meadows and other established herbaceous vegetation. However, a small number of undesirable perennial weeds may become established and persist within a sown meadow. Control of these perennial weeds can usually be achieved by good management rather than by the use of chemicals.

Docks (*Rumex crispus* and *Rumex obtusifolius*)

Curled Dock (*Rumex crispus*) and Broad-leaved Dock (*Rumex obtusifolius*) are biennial weeds of grasslands. The key to control is to cut in early to mid-June at flowering, just before seed set, preventing seed return to the soil. The timing may be judged by rubbing flowers between thumb and forefinger to check for the absence of seeds. If it feels gritty, seed has formed and it is too late to cut without removing the material. Scythe at the base or patch top, and if the timing is right it is not necessary to remove the cut

Scything Docks just before seed formation.

material and the Dock plant will complete its biennial life cycle and die. If the cut is too early the plant will perennate and survive to the following year (Evert et al., 2020). If cut too late, the cut material will contain seed and the cut material is best removed.

Germination and establishment of *Rumex obtusifolius* seed left on the ground is facilitated and enhanced by water-soluble biochemicals released by the leaves of mother plants, hence encouraging the spread and patch formation of Docks within a sward (Ohsaki et al., 2020). Long-term control of Dock species depends on consistent cutting at flowering every year to deplete the soil seed bank. One season of heavy return of Dock seed to the ground will set this process back many years.

Spear Thistle (*Cirsium vulgare*).

Creeping Thistle (*Cirsium arvense*).

Thistles (*Cirsium arvense* and *Cirsium vulgare*)

Creeping Thistle (*Cirsium arvense*) is a rhizomatous perennial, and Spear Thistle (*Cirsium vulgare*) is a biennial. Topping or scything at flowering will weaken Creeping Thistle, but will not kill it. There is no need to remove the cut material of Creeping Thistle as it will not contain, or go on to produce much seed during drying. Despite copious thistledown, Creeping Thistle relies heavily on rhizomatous vegetative means of spread, and much less so on seed production and dispersal. In contrast, Spear Thistle relies entirely on seed dispersal for colonising new spots, and the objective of cutting is to prevent seed formation. Again, top or scythe at flowering in late June to achieve this. At flowering Spear Thistle will have committed to a biennial life cycle and will die if cut at this time. For both species, an early cut in May or early June is counter productive as early-cut plants will re-sprout and go on to flower, although flowering date may be a little later than otherwise.

Creeping Thistle (*Cirsium arvense*) is present in most grassland as semi-dormant rhizomes, and cultivations of these grasslands by ploughing or harrowing can remove competition and release Creeping Thistle to dominate the disturbed ground. However, continued cultivation or topping will weaken the rhizomes and suppress Creeping Thistle.

Elimination of Spear Thistle (*Cirsium vulgare*) is fairly easy, as well-timed topping or scything will quickly deplete the soil seed bank, as will cutting through the crown with a hoe at just below soil level.

Common Ragwort (*Jacobea vulgaris*)

Ragwort is a late-June flowering biennial and there is no easy way of eradicating it (Defra, 2004). The only non-chemical method that works is to pull and remove the plants as soon as they are recognisable. Eventually, as numbers decline and

Ragwort (*Jacobea vulgaris*) is toxic to livestock, but does provide pollinator services.

Ragwort (*Jacobea vulgaris*) may be controlled by uprooting with a pronged Ragwort fork just before flowering, and removing plant material from the field.

Ragwort here being consumed by caterpillars of the Cinnabar Moth (*Tyria jacobaeae*).

the soil seed bank depletes, the task becomes easier. Topping grazed pasture, or cutting with the hay, is not an option because cut and dried Ragwort plants, on the ground or in hay, are as toxic to livestock as green living plants. The toxic alkaloids remain in the dried material.

Hoary Ragwort (*Jacobea erusifolia*) is probably as toxic as Common Ragwort and is frequent on high pH ground. It is similar to Common Ragwort, but a little later flowering and forms rhizomatous clumps. Hoary Ragwort can be pulled by hand, but invariably some rhizomes remain and these can re-sprout.

Hemlock (*Conium maculatum*)

Hemlock is a poisonous and unwelcome biennial of waysides. It is best cut at flowering by flail or scythe with the aim of preventing seed formation and eventually depleting the soil seed bank; this may take five to ten years or so.

Stinging Nettle (*Urtica dioica*)

Stinging Nettle is a rhizomatous perennial of open and semi-shaded situations on richer soils (Taylor, 2009). It is avoided by livestock in spring and summer, but often grazed in autumn. In pasture and meadows it may be suppressed by patch topping or scything. Eradication is very difficult and not always desirable. Stinging Nettle is often best left as a food plant for invertebrates, unless its presence is hindering grazing.

Stinging Nettle (*Urtica dioica*).

CHAPTER 15

HABITAT-SPECIFIC MANAGEMENT WITH EXAMPLES

Different habitats require different management, and in this chapter we describe the management of the different habitat types, illustrated with project examples.

Annuals

Communities of annual species may be maintained from year to year by a late summer/autumn cut followed by surface cultivation. Use a hand hoe, a rotovator, a power harrow, a tined harrow or disc harrows to cultivate the surface. By late summer most annual species will have completed their life cycle and shed seed, therefore do not plough as this will bury the seed. The purpose of the cultivation is to kill unwanted biennial and perennial weeds that may have established, and to prepare a seedbed for the next generation of annual species. Repeat the cultivation if necessary to rid the ground of perennial weeds, and after the final cultivation, roll in order to firm and level the surface.

The timing of the final cultivation will influence the composition of the mixture of annual species arising from the self-sown seed bank. Winter annuals will dominate from an early autumn final cultivation, and spring annuals will dominate from a spring final cultivation. Corn Poppy (*Papaver rhoeas*) is a winter annual, with seeds that germinate best and in profusion in early autumn, and the resulting seedlings are very winter hardy. Some germination of Poppy seed can occur in the spring, but far less than in the autumn. Corn Marigold (*Glebionis segetum*) is a spring annual, it is not winter hardy, and the seeds normally germinate in the spring. Peak flowering of a mixture of native annual cornfield species is usually around mid-June, provided that the preceding final cultivation is in autumn, winter or mid-spring. A later final cultivation will shift peak flowering from mid-June to later in the year by an amount roughly equal to the delay in final cultivation after mid-spring.

The above will be familiar to the grower of organic cereals, which, in the absence of herbicides, support abundant arable weeds, from both autumn and spring drillings. Some of these arable weeds are quite rare these days, and important for the ecosystem services they can provide (Storkey and Neve, 2018). Organic cereal crops are often a source of seed of interesting arable weeds that have sprung

Cornfield annual mixture sown in a garden setting.

up as volunteers from the soil seed bank. In fact, most Corn Poppy (*Papaver rhoeas*) seed is obtained in this way, usually as cleanings from combined organic cereals.

Meadow

The annual meadow cycle starts at the dawn of each year. As the days lengthen, livestock that are grazing the aftermath are pulled off, with the last leaving on 31 March. As the ground dries the opportunity may be taken to roll to provide a firm level surface for when the mower comes. Roll the meadow up to the first week in April (but no later), when the ground is sufficiently dry not to clog the roll. A Cambridge roll (ribbed roll) is better than a flat roll as it creates micro-niches for seed germination, invertebrates and so on. From 1 April the meadow is left to grow, and the first flowers to bloom are Cowslip (*Primula veris*) and Green Winged Orchid (*Anacamptis morio*) in April, followed by Dandelion (*Taraxacum officinale*) and Bulbous Buttercup (*Ranunculus bulbosus*) in May; these are then followed by many more species through June (this being the peak flowering month for lowland meadows) and into July. When the mower comes in July, all is cleared away exposing the shorn stubble. This greens up and the grazing livestock are returned to nibble off the fresh aftermath from then, off and on, through to 31 March of the following year. This is the meadow cycle, and year-to-year continuity of the cycle ensures the continued well-being of these biodiverse communities (Valerio et al., 2022).

Not all meadows are in a field. They can be created on any suitable patch of ground, in a park, a garden, on a village green, green space in residential, commercial and retail developments, on road verges, beside railway tracks and canals, leisure facilities and airports, on roofs and façades, in pots and planters. In most of these situations, haymaking and aftermath grazing is not an option, but cut and collect by various methods often is. Some sites will be best machine cut, while for others a scythe may be a better option. In this way (by annual cut and removal in mid-summer), an amenity, wayside or urban meadow can be maintained indefinitely.

The species content of a created meadow often follows a common pattern of change in the formative years, with species diversity (measured as species in flower) increasing

The meadow at King's College (Cambridge) was created in 2019 on a portion of the iconic Back Lawn. A seedbed was prepared and a mixture of annual species and perennial native meadow species was sown. This project has created a much-loved urban meadow, viewed here in year three with Moon Daisy (*Leucanthemum vulgare*) temporarily dominant. The meadow has been much studied and has been shown to provide substantial benefit in terms of biodiversity, ecosystem services, climate change mitigation, cultural gain and lowered maintenance costs, when compared with the adjacent and retained part of the original amenity lawn (Marshall et al., 2023).

exponentially in years nought to ten, then tailing off in years ten to twenty, then entering a period of stability, provided the management and the environment remain favourable. Typically, the first species to flower is Yellow Rattle (*Rhinanthus minor*) in years one and two, followed by short life-cycle perennials such as Red Clover (*Trifolium pratense*) and Moon Daisy (*Leucanthemum vulgare*) in years three and four, followed by medium-length life-cycle perennials such as Cowslip (*Primula veris*) and Knapweeds (*Centaurea sp*) in years five and six, and eventually by the very slow to mature perennials such as Quaking Grass (*Briza media*) and Orchids in years seven, eight and nine.

Succession in a created meadow will also be influenced by the interactions between individuals and between species, the interactions between these and the environment, and the multitude of interactions between interactions. Some common patterns emerge. Looking at meadow restoration, where existing species-poor grassland is scarified and over-sown with species-rich brush-harvested seeds, the main driver of change is often Yellow Rattle (*Rhinanthus minor*). Typically, Yellow Rattle will boom in year two, and this has a major influence on the survival of both the existing species and the introduced species.

When the created meadow was visited again in 2022 the meadow community had matured and had become more species diverse. Yellow Rattle (*Rhinanthus minor*) was less frequent and the diverse community was dominated by Quaking Grass (*Briza media*), Upright Brome (*Bromopsis erecta*), Bird's-foot Trefoil (*Lotus corniculatus*), Fairy Flax (*Linum catharticum*), and many other species. Astonishingly, thousands of Pyramidal Orchids (*Anacamptis pyramidalis*) and a few Bee Orchids (*Ophrys apifera*) had colonised the 8ha (20-acre) field.

Meadow Creation in the Cotswolds

Meadow creation: this site in the Cotswolds is a former arable field on calcareous soil. A seedbed was prepared following the last arable crop, and the field was sown in 2010 with brush-harvested seed from a nearby species-rich calcareous meadow (*see* image overleaf). Establishment was good, and when visited in 2010, the created meadow was dominated by Yellow Rattle (*Rhinanthus minor*), with lesser amounts of the faster establishing native legumes. Managed under environmental stewardship, the created meadow was cut for hay each year on or after 15 July and the aftermath grazed.

Meadow Restoration in the Cotswolds

The images that follow illustrate the changes in vegetation cover that have occurred when semi-improved agricultural grasslands on calcareous ground are scarified and over-seeded with brush-harvested seed from a nearby species-diverse calcareous meadow.

Meadow restoration: this is the donor site for the brush-harvested seed used in the restoration of the semi-improved grassland fields seen overleaf, these each being photographed and surveyed in successive years.

Year nought: The semi-improved vegetation of the recipient fields was typically dominated by Rough-stalked Meadow Grass (*Poa trivialis*), Yorkshire Fog (*Holcus lanatus*) and Perennial Ryegrass (*Lolium perenne*). These fields were prepared for overseeding by cutting and baling the forage in the summer, followed by scarification of the ground. The donor seed was applied in the autumn by broadcasting, and grazing was resumed after seeding to tread in the sown seed and to suppress competition from the existing grasses.

Year two: The second full year after overseeding is characterised by a population explosion of second-generation Yellow Rattle (*Rhinanthus minor*), and an associated decline in populations of Yorkshire Fog, Rough-stalked Meadow Grass and Perennial Ryegrass.

Year one: Yellow Rattle (*Rhinanthus minor*) has established at moderate density in the first full year after seeding, as is normal.

Year four: Many of the early establishing perennial meadow species that have been sown are flowering by year four, including Moon Daisy (*Leucanthmum vulgare*), Common Knapweed (*Centaurea nigra*) and Rough Hawkbit (*Leontodon hispidus*).

From time to time, individual species in a meadow may appear to become over-dominant. This is a cyclic feature and will be most obvious in showy species such as Moon Daisy (*Leucanthemum vulgare*), Hogweed (*Heracleum sphondylium*) and Wild Carrot (*Daucus carota*). No specific action is required. As sure as night follows day, a bust will follow a boom (Flory et al., 2017) if nature is allowed to have its way.

The length of the boom/bust cycle is related primarily to the length of the life cycle of a species. Field observations suggest that typical boom/bust cycles may vary in length from eight to twenty years.

- Yellow Rattle (*Rhinanthus minor*): there is a population explosion in years two to three after introduction. Cycle length is eight years (eight generations).
- Moon Daisy (*Leucanthemum vulgare*): there is a population explosion in years three to four after introduction. Cycle length is eight years (four generations).

Year six: Slower to establish perennial meadow species, such as Bird's-foot Trefoil (*Lotus corniculatus*), Quaking Grass (*Briza media*) and Cowslip (*Primula veris*), are flowering by year six.

Year eight: The restored calcareous meadows are approaching maturity by this time, with, in patches, hundreds of Common Spotted Orchid (*Dactylorhiza fuchsii*) flowering beneath a thin canopy of Upright Brome (*Bromopsis arecta*). In following years, the Common Spotted Orchids produced a second-generation population explosion, and then later declined.

- Common Knapweed (*Centaurea nigra*): population explosion in years eight to ten after introduction. Cycle length is eight to ten years (three generations).
- Cowslip (*Primula veris*): population explosion in years ten to fourteen after introduction. Cycle length is ten to twenty years (three to five generations).

In a large hay meadow, species at different locations will probably be at different stages of their population explosion and subsequent decline, giving rise to a dynamic patch-like distribution of changing communities across a field. Many other factors influence the structure and distribution of community patches across a meadow, too many to be considered here in detail. Meadow rings and circles will serve as examples.

Yellow Rattle (*Rhinanthus minor*) can grow in ring formation. An isolated plant may shed seeds to a distance of, say, 0.25m (0.8ft). The resultant plants do the same, as do subsequent generations, and so the patch grows outwards at around 0.25m/year. While this is happening, the flora of the central area of the patch becomes dominated by Yellow Rattle tolerant species and depleted of Yellow Rattle host species. Thus an advancing ring of Yellow Rattle is formed. This phenomenon is noticeable in extensively grazed pasture, but less so in meadows because at hay-making, seeds are distributed randomly up to and beyond 0.25m/year, blurring formations.

Fairy rings are another example of this type of growth (Shantz and Piemeisel, 1917), important for increasing above-ground plant community structural diversity and species diversity (Zotti et al., 2020; Johnson and Gilbert, 2015). As the fairy-ring fungus grows out from a point of origin, the older hyphae die and the younger hyphae advance further outwards, forming a ring visible above ground as fruiting bodies and/or a ring of darker green vegetation. Sometimes the ring may be visible as differences in the flowering plants that grow, or don't grow, on the ring, in the ring, and outside the ring.

A meadow full of Cowslips (*Primula veris*) is glorious for a few weeks in spring, before the other yellow spring meadow flowers bloom en masse. Sometimes, with luck, Cowslips may appear in ring formations, possibly as a consequence of a complex interaction with below-ground ring-forming fungi. As can be seen in the illustration, Cowslip is absent from the ring, as is Hogweed (*Heracleum sphondylium*) on close inspection, while both species have recolonised (or revived) within the ring. In this instance, other common meadow species seem not to have reacted in this way (MacIntyre, 2006).

Meadow circles are formed when a rhizomatous (or stoloniferous) perennial species grows out vegetatively from a point of origin and forms a circular or irregular patch, getting larger each year. This is a feature of clonal species such as Mouse-ear Hawkweed (*Pilosella officinarum*), Bugle (*Ajuga reptans*) and Glaucous Sedge (*Carex flacca*). Glaucous Sedge is a common indicator species of old species-rich hay meadows, and is an excellent example, at first forming circles expanding outwards. As these circles grow the centre dies and the circle becomes a clonal ring, often

Table 3: Meadow Restoration in the Cotswolds

Changes in species abundance following scarification of species-poor grassland and over-seeding with brush-harvested seed from a nearby species-rich calcareous meadow

Species	Year 0	% sown	Year 1	Year 2	Year 4	Year 6	Year 8
Agrostis stolonifera	3	0	2	1	0	0	0
Anthoxanthum odoratum	3	8	4	5	3	3	3
Arrhenatherum elatius	3	1	3	2	1	1	1
Briza media	0	10	0	0	1	2	2
Bromopsis erecta	0	15	0	0	2	3	3
Centaurea nigra	0	2	0	0	3	2	2
Centaurea scabiosa	0	1	0	0	1	1	1
Cirsium arvensis	2	0	2	2	1	1	1
Crepis capillaris	0	1	0	1	2	2	2
Cynosurus cristatus	2	10	2	3	3	3	3
Dactylis glomerata	4	2	3	2	1	1	1
Dactylorhiza fuchsii	0	?	0	0	2	3	3
Festuca rubra	2	6	2	3	3	3	3
Heracleum sphondylium	0	2	0	2	2	2	2
Holcus lanatus	4	2	3	1	1	1	1
Senecio jacobea	2	0	2	2	2	2	2
Knautia arvensis	0	1	0	0	1	1	1
Leontodon hispidus	0	3	0	1	2	2	2
Leucanthemum vulgare	0	1	0	2	2	3	2
Lolium perenne	3	1	2	0	0	0	0
Linum catharticum	0	3	0	1	2	2	2
Lotus corniculatus	0	2	0	0	3	3	3
Medicago lupulina	0	2	0	1	2	2	2
Ophioglossum vulgatum	0	?	0	0	2	3	3
Plantago lanceolata	2	4	3	5	3	3	3
Plantago media	0	1	0	1	2	2	2
Poa trivialis	4	0	3	1	0	0	0
Primula veris	0	2	0	0	3	3	3
Ranunculus acris	3	6	2	4	2	4	3
Rhinanthus minor	0	12	3	5	4	3	2
Trifolium pratense	0	2	0	1	2	2	2
Total species	**13**	**28**	**14**	**22**	**29**	**29**	**29**

The abundance of species is averaged over seven fields seeded in different years and estimated on a scale of 0–5 where 0 = absent, 1 = rare, 2 = occasional, 3 = frequent, 4 = abundant and 5 = dominant. Year 0 refers to the abundance of species before seeding. Percentage sown refers to the typical composition of the mixture sown expressed as a per cent seed number. Seeding was in the autumn of year 0. Years 1–8 refers to years after seeding, with abundance observed in June.

A ring in a Cowslip (*Primula veris*) meadow, probably the creation of mycorrhizal fungi.

many decades old. At SU101643 on Milk Hill at Pewsey Downs NNR sits a huge clonal ring of Glaucous Sedge. The ring grows in calcareous grassland on a steep southwest-facing scarp. When seen on 9 May 2005 the ring had a width of around 1m (3ft), and the circle it made had a diameter of 19m (62ft) to its outer edge. The ring was complete except at its northwest edge where it had failed to pass through a group of active Badger setts.

It is an outward sign of a healthy ecosystem if nature alone can arrange meadow flowers in delightful patches, rings and circles. It is a far cry from chemical mono-cropping in straight rows and neatly ordered gardens.

All this aside, the meadow must be mown, although not all is lost if it is not. A meadow can recover from the occasional missed season, but if cut more frequently through the year, or substantially less frequently than once every year, it turns gradually into pasture or tussock, both of which can be valuable and interesting ecosystems in their own right.

Pasture

On the farm, pasture is ground that is grazed on and off through the year and is not cut for hay. On a commercial farm pasture will most frequently consist of sown Perennial Ryegrass (*Lolium perenne*) leys, producing high forage yields in response to the application of fertiliser and herbicides – but it need not always be this way. Before the agricultural revolution grazing pasture was almost as biodiverse as were meadows at that time. Biodiverse pasture will always yield less forage than the modern Ryegrass ley. If the value of the ecosystem services are added in, native biodiverse pasture is far more productive than the modern ley.

Wild pasture can be restored and created using the same methods as may be used for restoring and creating meadows. Yellow Rattle (*Rhinanthus minor*) can be used to aid the restoration process, taking care not to cut or graze it until the restoration process is complete. Species selection will be a little different to that for a meadow because pasture management favours grazing-tolerant species with a rosette or creeping habit, rather than the more erect habit of meadow species. This distinction becomes rather less critical as pasture is more extensively grazed, allowing many typical meadow species the opportunity to survive as grazing pressure relaxes.

Yellow Rattle does not normally survive in pasture, nor does Red Clover (*Trifolium pratense*), as both have an upright growth habit and both are grazing intolerant. White Clover (*Trifolium repens*) normally replaces Red Clover when a meadow is converted to pasture.

Management of pasture with livestock can be by mob grazing or extensive grazing, or a combination of the two. With mob grazing, the vegetation in a compartment is grazed down at high stocking rates, the livestock are moved on, the residue vegetation is topped and the compartment is rested. The careful timing of the rest period allows selected groups of early, mid- or late season species to flower and seed, if this is desired. Extensive grazing consists of low stocking rates on open areas of rough pasture, allowing moderate flowering through the season. Any topping would be restricted to patches targeting any specific weed problems.

Species-rich pasture, surviving here because the ground is too steep for a tractor, but can be maintained by extensive grazing.

In a garden or park or on other amenity grassland, the equivalent to wild pasture would be the 'flowering lawn' (similar to extensively grazed pasture), or the 'no mow May lawn' (similar to mob-grazed pasture at rest). The method of establishment would be the same as for restoration or creation of a meadow, and the seed used may be a meadow mixture (with or without adapting the species content to include more creeping and rosette species). Once established, this species-rich amenity grassland may be managed at a much reduced cutting intensity, but without leaving the vegetation to grow tall into a hay crop. The timing and the height of the cuts can be chosen to suit specific objectives, taking into account the species present and the prevailing conditions. Overall, the outcome will be short vegetation full of wild flowers, appearing somewhat midway between mown amenity grass and a meadow in full bloom.

Wayside and Tussock

Wayside vegetation is the vegetation growing on the verges, banks and cuttings of roads, railway lines, canals, tracks and paths. Tussock vegetation is the highly structured, tall herbaceous vegetation that develops in the absence of management and grazing. Both wayside and tussock benefit from minimal management, and both deliver a whole host of ecosystem services.

Species-diverse wayside vegetation may be established by using meadow restoration and creation methods. If the existing wayside vegetation consists of native grasses, but is species poor, the growth may be cut back, and the ground scarified and autumn sown with a mixture containing Yellow Rattle, thus restoring the vegetation to a more species-diverse community of native species. Often, engineering works will create associated verges, banks and cuttings consisting of bare soil. If intending to create native wild vegetation, never finish the earthworks with topsoil, but sow directly onto the local subsoil brought up by the engineering works. If surplus topsoil has been created by the works and cannot be used elsewhere, it is best buried at depth on the site. The lower the nutrient status of the sown substrate, the more species diverse will be the

A roadside bank, not mown until late summer so as to let the meadow wild flowers bloom.

The Baldock Bypass (opened in 2006) created 105ha (260 acres) of new additional chalk grassland along a stretch of 6km (3.7 miles) of new road. A relic 0.2ha (0.5 acre) chalk grassland community on nearby Weston Hill was used as a source of stock seed for propagation off-site of the seed of the native species required for sowing along the route. Most of the seeded landforms consisted of pure chalk, and seed was applied by hydraseeding (Baldock Bypass, 2009). The images are of the entrance to the road tunnel in 2006 and in 2023, by which time the bare chalk cutting had become well vegetated with native species, including orchid colonists.

A closer look at Baldock Bypass in 2023 shows that the native chalk-grassland species have established well on the former bare chalk landform. Because of the low nutrient status of the substrate, the maintenance requirement for this created chalk grassland is no more than occasional scrub control.

Batheaston Bypass cut deep through hillside limestone subsoils, soil that was used to create high-shielding embankments where the bypass crossed Bathampton Meadow. These embankments were sown with a mixture of native limestone-loving species, stock for many of which was collected locally and grown off-site to produce the required seed. Pictured here in 2010, the native ground cover is well established, and orchids are arriving as natural colonists.

resultant vegetation, and the easier will be the management of that vegetation. Species choice will be dictated by the soil and location conditions, and the choice may be a classic meadow mixture, or a mixture specific to the site, and may include a nurse of cornfield annuals.

The seed having been sown and rolled in, there would normally be no need for any immediate follow-up management apart from the spot control, after one year or more,

Odd Down Park and Ride was constructed on the limestone plateau south of Bath, and it was always intended to mix parking with nature. The first diggers on site removed the topsoil to expose the limestone brash subsoil. The brash provided the perfect substrate for sowing lime-loving native meadow species between and around parking spaces. This image from 2013 shows how well the sown species have established. Management consists of a single cut in the autumn to tidy the site.

Odd Down Park and Ride has matured in later years, and orchids have naturally colonised the site, as seen in this image from 2000 – as have butterflies also. The Park and Ride now has an established colony of the rare and declining Small Blue butterfly (*Cupido minimus*), with its caterpillars feeding on the sown Kidney Vetch (*Anthyllis vulneraria*). The Small Blue is the smallest of our resident butterflies.

of any injurious weeds that may become apparent. If desired, the established wayside vegetation may be managed as a meadow with an annual cut and collect in July or August. However, if the substrate fertility is low or very low, or if there are resource constraints, it may not be necessary or possible to cut and collect the wayside vegetation annually. In fact, most waysides are best not managed as meadows, but rather left unmanaged for the ecosystem service benefits that arise from undisturbed species-rich vegetation. However, without any management these waysides eventually revert to scrub, and succession towards woodland gathers pace. This natural succession can be halted and reversed by flail scrub bashing every two to ten years. Flailing is best restricted to the winter months when the vegetation is not in full growth and when birds are not nesting.

While leaving the wayside wild and largely unmanaged, it is important for road safety to maintain sight lines as always cut, and on major routes, for appearance, to maintain a cut edge to the verge. On minor routes the cutting of verges can be left to the autumn.

Tussock vegetation consists of robust perennial species with an upright growth habit and an ability to survive in an unmanaged and ungrazed grassland setting. Example species are Knapweeds (*Centaurea sp*), Meadow Crane's Bill (*Geranium pratense*) and Field Scabious (*Knautia arvensis*). Tussock vegetation may be established using meadow restoration and creation methods, choosing species that will create the desired tussock-like structure. The resultant community will not be as species diverse as a meadow, and it will be a little less flowery in summer. However, tussock vegetation provides a welcoming habitat for creatures throughout the year: ground-nesting birds in spring, pollinators in summer, wide-ranging fauna in autumn foraging for food, and invertebrates, mammals and raptors, hibernating and hunting over winter. It may look a little unkempt in winter, but will be hugely popular with wildlife, and will always look better if the edges and pathways through it are kept neatly mown. It is an ideal ground cover for an orchard, with fallen apples attracting even more fauna. Like wayside vegetation, completely unmanaged tussock vegetation will eventually revert to scrub and start on the pathway to woodland. Too much scrub may be dealt with by winter flailing in patches, but not every patch every year. At most, flail every other year, but leave the interval for longer if possible.

Floodplain, Wetland and Water Edge

Wetland and water-edge communities include floodplain meadows, wet meadows, wet pasture and water-edge vegetation. They can all be restored by the scarification method and sowing native seeds, including Yellow Rattle, in the autumn. Sowing complete mixtures of native species onto bare prepared soil can also create wetlands.

Floodplain meadows are characterised by MG4 grasslands (Rothero et al., 2016). These communities are adapted, as of necessity, to extremes of hydrology, often being inundated by floodwater in winter as river levels rise, and often subject to extremes of drought in summer, all depending on the weather. It is a difficult community to create or restore, and even on fields that are ploughed up, former MG4 situated adjacent to existing MG4, the restoration progress is slow, and may take several decades. But it is possible, and eventual success depends on the mid-summer hay cut and removal followed by well-managed aftermath grazing. It also

Uncut and ungrazed tussock vegetation in winter provides a hibernation habitat for invertebrates and other creatures, foraging for small mammals and birds, and a hunting ground for raptors.

Streamside vegetation with Pendulous Sedge (*Carex pendula*) and Harts-tongue Fern (*Asplenium scolopendrium*).

depends on moderately stable hydrology. Upstream development run-off combined with increasingly extreme weather events increases the inundation of floodplain meadows, and frequent and prolonged inundation is lethal to some component species of floodplain meadow grasslands, setting back the restoration process towards MG4 by several years at a time. Increased inundation also increases nutrient deposition, slowing progress towards full restoration (Wagner et al., 2020).

If the frequency of extreme inundation events exceeds the length of the recovery period, the restoration of MG4 grassland can go into reverse. This is probably one reason why restoration of floodplain meadows is so difficult and slow. Other reasons may include the time taken to build the complexities of the interactions between slow-to-establish and long-lived MG4 species (as most are), such as Great Burnet (*Sanguisorba officinalis*), Meadowsweet (*Filipendula ulmaria*), Saw-wort (*Serratula tinctoria*) and many others. Another feature of floodplain meadows is that flooding brings in seeds from sources upstream, and some floodplain meadow species have acquired adaptations to this mode of seed dispersal. The seeds of Snake's-head Fritillary (*Fritillaria meleagris*) are flattened and triangular in shape with small air sacks on the outer edge, and they float, as do the seeds of Yellow Flag-iris (*Iris pseudacorus*).

The hydrology of wet meadows is different, with wetness coming from combinations of precipitation, springs, high water tables and poor drainage, and fluctuations in wetness throughout the year are less extreme than on floodplain meadows. These wet meadows are slow to dry out in summer and are usually mown later than floodplain meadows. Floodplain meadows are usually ready for the hay cut in early to mid July, while wet meadows may not be ready till late July or into August.

The Bathampton ox-bow lake and associated native vegetation was created during the construction of the Batheaston Bypass (1995–97), and is situated on a triangular-shaped 9ha (22-acre) site between the River Avon and the bypass, on the edge of Bath (Gosney et al., 1997). The site is dual purpose. The excavation and removal of 84,000cu m (297,000cu ft) of soil provides water-storage capacity during flooding of the River Avon, and the construction and seeding of a diversity of land forms has created a nature reserve rich in rare native plants and other life forms. The site is managed by the Avon Wildlife Trust, employing occasional extensive cattle grazing, and I thank Ann Skinner for providing additional information on this project.

Heron Meadow, Nene Park, Peterborough, sown in 2014 using a mixture of species characteristic of floodplain meadows. The 2ha (5-acre) meadow was created by broadcasting the seed onto prepared bare soil, and is pictured here in June 2022 buzzing with pollinators and much appreciated by visitors.

Wetland on rough ground, or ground that does not dry out, may be too uneven to mow for hay or may be too wet to carry a tractor. Here we include open ground as well as ditches, marsh and bogs, and the margins of ponds, watercourses and lakes. The restored or created vegetation for these habitat types is best left unmanaged after seeding, apart from the occasional control of unwanted injurious weeds, scrub and developing woodland. Extensive grazing with cattle will help maintain species diversity on dryer patches of grassland, and slow the succession to scrub and woodland.

In an urban, garden, park or landscaped setting, management of wetland and water-edge vegetation is best kept to a minimum. Hay making and grazing are usually not

The edges of ponds provide perfect conditions for native emergent species. Here, Purple Loosestrife (*Lythrum salicaria*) and Great Willowherb (*Epilobium hirsutum*) have their roots in the water, playing a beneficial role in the wetland ecosystem.

possible. Simply delight in the uncontrolled wildness of the vegetation, while keeping edges and pathways neatly mown. Do not remove dead standing vegetation, such as that of Wild Angelica (*Angelica sylvestris*) and other native species, in autumn, as seeds provide winter nourishment for birds, and hollow stems a secure hibernation spot for invertebrates.

Hedgerow and Scrub

Hedgerow ground flora can be created and established prior to hedge planting, or may be sown onto bare soil at the same time as planting a hedge, whilst the degraded ground flora of an existing hedge may be added to and diversified. Creation and restoration methods will be the same as those used for meadows, while the roots of an existing hedge and any weeds associated with the hedge will make scarification and ground preparation a little more difficult than on open ground. Nevertheless, gradation of the ecological conditions from the centre of the hedge to the edge of its influence (an ecocline) creates multiple niches and opportunities for high species diversity.

Hedges are important for wildlife, and their management has a huge influence on their value to nature. Hedgerows laden with berries are alive in the autumn and winter with an avian feeding frenzy. While some hedgerow climbing species produce flowers and fruit on the current year's growth, as do Brambles (*Rubus fructicosa*) and Black Nightshade (*Solanum nigrum*), most hedgerow shrubs produce flowers and fruit on the previous year's growth. This is so for Hawthorn (*Crataegus monogyna*), Dogwood (*Cornus sanguinea*), Blackthorn (*Prunus spinosa*) and Spindle (*Euonymus europaeus*), among many others. Therefore hedges that are flail cut every year in the autumn, removing the current year's growth, will not produce any flowers or fruit on the hedge in each following year, except for a few climbers.

For a hedge to produce its full yield of fruit it must not be flailed in the previous year. And to ensure that the hedge fruit remains available for the autumn avian feeding frenzy, flailing in the current year must not be before hedges are stripped of most of their fruit, usually sometime around mid-winter. Spindle (*Euonymus europaeus*), Holly (*Ilex europeus*) and Ivy (*Hedera helix*) berries mature a little later. From mid-winter to the start of the bird-nesting season in

Hedgerow bounty: late pollen on Ivy (*Hedera helix*) and juicy red berries on Black Bryony (*Tamus communis*), food for pollinators and birds. If hedges must be flailed, it is best left to midwinter.

Hedge laying is a nature-friendly alternative to flailing.

March, there is a narrow, nature-friendly winter window for hedge flailing. This is all right when cutting from the highway, but can only be done from the field side when the ground is sufficiently dry.

Most farmland hedges are flailed in September and October when the ground is dry. If this is done on a one- or two-year cycle there will be no autumn fruit source for birds at any time. If it is done on a three-year cycle, hedgerow berries will only be available for one year in three. The benefit to nature increases as the length of the cutting cycle increases beyond three years, but it also gets more difficult to flail and produce a neat hedge. Alternatively, hedge laying, which is hard, skilled work, can be done on a cycle of around ten years. Likewise hedge coppicing, which entails cutting the hedge to the ground, can also be done on a cycle of around ten years. Coppicing rejuvenates a hedge, it does not kill it, and provided the coppiced hedge is fenced, it will regrow quickly. Long-cycle laying and coppicing are better for hedges and wildlife than regular flailing. Coppicing and laying also allow an opportunity to select and spare saplings, allowing them to grow on to become hedgerow trees managed as standards or pollards.

Sheep will eat out and kill a hedge unless it is stock fenced on both sides. Cattle and equines will not kill a hedge, and fencing need not be on both sides. Sheep will also eat out the herbaceous vegetation at the base of a hedge. If not grazing with sheep, the restored/created species-rich vegetation at the base of a hedge may be maintained with a single cut taken at the same time as cutting the hedge. If flailing, this basal ground cut should be in winter, and the width of the cut may be 1m (3ft) or more, while allowing the hedge to be at least 2m (7ft) wide.

In the absence of management, Blackthorn (*Prunus spinosa*) will expand sideways from the hedge and encroach on the field. An unmanaged hedge will also extend vertically, with trees gradually replacing bushes, becoming in time a woodland strip or a corner spinney. Often, the hedges between holdings will be less well managed than those within, not usually because of a lack of ownership, but because both sides will often claim ownership, as is natural, and cannot agree.

Hedges and scrub are similar in the native plant species they support, and a feature of both is high ecological niche diversity and the ability to support high species diversity. Scrub is a patch feature rather than a linear feature, and is a passing phase in the natural and inevitable transition of unmanaged grassland to woodland. Scrub is a consequence where there is no cutting of the grassland, and where there is only low-intensity grazing, or none. Stages in the transition to woodland might consist of first, tussock grassland; then grassland with expanding browse-resistant Bramble (*Rubus fruticosus*) patches; then bramble patches that are colonised by saplings of hawthorn, blackthorn, ash and oak; then patches that progressively meet up; and finally, the canopy of the tree species meets overhead, shading out the remnants of the grassland and the brambles.

Scrub is potentially more species diverse than woodland or pasture, and the objective of positive scrub management is to maintain an intermediate patchwork phase that consists of approximately equal parts ground cover of scrub and tussock grassland. This may be achieved by selective and periodic partial flailing of the scrub on a five- to ten-year cycle.

An orchard shares features with hedgerows, scrub and woodland, and can be highly biodiverse if allowed to run a little wild. A ground flora consisting of a meadow is good,

Scrub, in this case Hawthorn (*Cretagous monogyna*) and Brambles (*Rubus fructicosa*), can add welcome structural diversity to species-rich pasture.

and is an option that fits in well with managing the orchard trees. The meadow will be flowering in spring and early summer while the orchard trees are flowering and setting fruit. After the meadow is cut, access will be easy for getting in to harvest fruit, and provided the trees are guarded against browsing, the orchard can be grazed in the autumn and winter. Sheep are best for this as their browse height is limited. This is all fine, though making hay in an orchard is not easy, as the trees get in the way. An alternative is to manage the orchard ground flora as tussock grassland with some autumn/winter grazing, and topping every two to three years. Allowing fallen fruit to remain on the ground will boost numbers of supported fauna.

Woodland

When planting up a new wood, select native tree species of local origin that are characteristic of and suited to the location – or in preference, simply allow woodland to develop by natural regeneration. Natural regeneration can be surprisingly quick if established woodland is adjacent or close by. A canopy of pioneer woodland species such as Goat Willow (*Salix caprea*), Elder (*Sambucus nigra*) and Hawthorn (*Cretagous monogyna*) can develop naturally in fifteen years. This will then proceed, over the coming decades, to develop into secondary mature native woodland, a process that can be speeded up by broadcasting, or pressing in, native tree seeds of Oak (*Quercus sp*), Ash (*Fraxinus excelsior*), Field Maple (*Acer campestre*), Hazel (*Corylus avelana*), and others. If restoring native ground flora to an existing wood it is good to first remove any alien or ornamental tree species that may be present.

Woodland ground flora in spring, benefitting from an open tree structure and moderate light penetration.

Woodland ground flora cannot be established at the same time as planting trees because most woodland floor species have poor tolerance of direct sunlight and poor competitive ability in the open. Tussock grassland or a meadow are a good companion for young trees, and as the canopy slowly develops, these open ground communities will die back, as will most weed species. Bracken, Brambles and Stinging Nettles are species of woodland edge and clearings, and will not tolerate a closed woodland canopy, naturally declining as the wood matures. Ground preparation for establishing woodland ground flora consists of raking off leaf litter and surface scarification in the autumn or winter. Below-surface cultivation is not an option because of the presence of tree roots. Seed may be surface sown and rolled in, in the usual way.

Woodland is the climax wild vegetation over most of Britain and Ireland, in the absence of grazing, cutting and

Woodland ground flora consisting mainly of Bluebell (*Hyacinthoides non-scripta*) and Yellow Archangel (*Lamiastrum galeobdolon*), both benefitting from a coppice cycle and moderate light penetration.

cultivation. Provided the wood is fenced, other management is not necessary for the survival of woodland. Without intervention, woodland develops a natural layered structure consisting of tall trees, shrub layer and herbaceous layer, and this is interrupted spatially by fallen or standing dead trees, triggering a cyclic patchwork of death, renewal and growth across layers. Dead wood and upended root plates are so important to nature, and should be left where they stand or fall, if it is safe to do so. There is more life in a dead tree than in a living tree. Coppiced Hazel with Ash and Oak standards is a common form of traditional managed woodland, yielding timber and fuel, as well as being of huge benefit to nature. Coppice woodland may be left unmanaged and allowed to return to nature, or a coppice cycle may be restored. Both options are better than grubbing out.

THIS PAGE IMAGES: It is often said that there is more life in a dead tree than in a living one. The diversity of fungi growing on fallen trees is astonishing.

Grazing deer may harm woodland biodiversity. There are two native species and four introduced species of deer, all of which selectively browse regrowth of woody species and selectively graze favoured species of woodland ground flora, reducing diversity. Other harm may be done by Rhododendron (*Rhododendron ponticum*), an invasive alien that may replace the shrub layer in woodland and shade out the ground flora. Further, it is well known that the alien Grey Squirrel (*Sciurus carolinensis*) has ousted the native Red Squirrel (*Sciurus vulgaris*) from woods across most of England.

Orchids

There are around fifty-one species of Orchid native to Britain and Ireland, around ten species of which may be sufficiently common and widespread to be considered for inclusion in the restoration or creation of meadows and other native herbaceous vegetation, within their native range.

Our native Orchid species are perennial and belong to the Orchid family, the Orchidaceae. They are distinguished from other plants by having parallel leaf venation, below-ground paired tubers or rhizomes, an inferior ovary, flower parts in threes (trimerous) and flowers borne individually facing outwards on an unbranched spike. The three sepals of each flower are usually equal and petaloid in shape and colour. The three petals are usually unequal, the uppermost usually highly differentiated to form a labellum that is then positioned lowermost by a 180^0 twist in the subtending pedicel of each flower. The labellum is usually the most obvious and developed feature of the Orchid flower and is often extended to the rear to form a nectar-containing spur. Pollen is contained in two to four detachable pollinia, each usually on a filament provided with a sticky pad (viscidium) at the base. Some species of Orchid are primarily outbreeding and some are primarily inbreeding (by self-pollination). Hybridisation between some species is frequent, and within species variation can be considerable.

All Orchids have dust seed, each seed consisting of a few cells, and each flower on an inflorescence spike producing around 1,000 seeds. The life cycle strategy of Orchids centres on producing huge numbers of very light seeds that can disperse on the wind. Each seed is the result of a fertilisation event, and therefore for successful pod fill, each ovule must meet up with a pollen grain, and each flower will require at least as many pollen grains as ovules that it produces. This is the reason for packing huge numbers of pollen grains into pollinia, so that they can be passed intact between plants during cross-pollination. It is also the reason for the elaborate diversification of flower forms between species. This is designed to attract dedicated species-specific pollinating invertebrates, thus reducing wastage of pollinia. The more specialised the association between Orchid species and its dedicated pollinator, the more vulnerable the association is to pollinator absence, and therefore some Orchid species have opted for self-pollination as a fall-back option (Bee Orchid, *Ophrys apifera*), or as an alternative obligate lifestyle (some Helleborines, *Epipactis sp*).

Take care to sow Orchid seed on a still day. It is good to mix the seed with a soil carrier, and even better if the soil contains Orchid mycorrhizal fungi. Select some likely looking existing companion vegetation, and broadcast the mixture of seed and soil on the intended spot. The seed will be washed down through the soil profile by rain. Orchid seeds carry few nutrient reserves, but can survive in the soil seed bank for five to ten years (Whigham et al., 2006). Putting seeds in nylon mesh packets, placing these in the soil for varying periods, and testing viability on retrieval, may test longevity in the soil. Unlike most wild species, Orchids do not require light to trigger germination, but they do require the presence of hyphae of their chosen special mycorrhizal fungal partner (Jacquemyn et al., 2013). These mycorrhizal fungi are fairly common, but are not ubiquitous in soils (McCormick and Jacquemyn, 2014). They may or may not be present in the absence of Orchids, but are sure to be present in the soil around existing Orchid plants, which often appear to cluster in mycorrhizal hotspots (McCormick et al., 2016; Waud et al., 2016).

Orchid seed can be germinated in the laboratory on nutrient agar, and will grow in the absence of mycorrhizal fungi if provided with a source of nutrition. In soil, the Orchid seed reacts to the presence of a hypha of its future partner mycorrhizal fungus by expanding a cell, and this cell is then entered by a single hypha, forming an enduring endo-mycorrhizal symbiosis. From that event on, for three to four years, the fungal partner feeds and inhabits the cells of the growing young Orchid, all below ground. It is not a known mutualism, because there appears to be no benefit to the fungus (Selosse, 2014). As the relationship ages, the Orchid becomes less choosy and hitches up with several other mycorrhizal species, and more as time goes by.

Eventually, after three to four years below ground, the young Orchid tubers produce shoots, and leaves expand above ground. For most species, vegetative rosettes appear

Bee Orchid (*Ophrys apifera*) growing between two plants of Common Broomrape (*Orobanche minor*), a parasitic species of a range of grassland host species, but not Orchids.

above ground in spring sporting unmarked, entire leaves, lanceolate or elliptic in shape (for example Pyramidal Orchid, *Anacamptis pyramidalis*; Greater Butterfly Orchid, *Platanthera chlorantha*), or with leaves marked with distinctive leopard-like dark spots or blotches (for example Common Spotted Orchid, *Dactylorhiza fuchsii*; Early Purple Orchid, *Orchis mascula*). A few native Orchids are wintergreen. Bee Orchid (*Ophrys apifera*) is first seen above ground in mid-October as distinctive spreading rosettes of hairless leaves that are at first green, turning grey/green in winter, at which time Bee Orchid is vulnerable to sheep grazing. These winter leaves senesce in late spring as the flower spike appears. In contrast, Autumn Lady's-tresses (*Spiranthes spiralis*) produces a basal rosette of rounded green leaves in mid-summer, before proceeding to flower in September.

From seed germination to flowering is six to eight years for most Orchid species. Management of mature Orchid populations will consist of ceasing to cut or graze from the end of March each year right through to the point at which flowering has finished, seed has set and pods have dehisced. For most grassland Orchids this will be around mid-August, rather later than the normal hay-cutting date. If a late hay cut can be taken, this will be beneficial. Otherwise, depending on the nature of the substrate and the vegetation, leave the vegetation unmanaged, or top/graze, or cut and remove after seed dispersal.

Seed set can be poor when the population size of both the Orchids and their pollinators is low. Non-swollen ovaries are a sign of poor seed set, and retained pollinia are a sign of lack of pollinator visits. Intervention is possible by gently placing a small finger or a pencil into the opening of a flower with retained pollinia. The sticky pads of the viscidium attach to the surface of the pencil or fingernail, and on withdrawal out come the pollinia attached by filaments to the viscidium. The pollinia are in a swept-back position at first, but within a minute or so the filaments curve to a swept-forward position, ready to effect pollination. At this point the pencil or finger may be inserted into an open flower on a different plant but of the same species, so that the pollinia come into contact with the stigma.

Most grassland Orchids are polycarpic and will continue to flower in successive years. The tubers (or rhizomes) can

Common Spotted Orchid (*Dactylorhiza fuchsii*) as a natural colonist in a restored meadow, probably having arrived from far off as seed floating on the breeze.

multiply below ground, sometimes producing a clump of clonal individuals, the number of which is a rough guide to the age of the plant. If the sown first-generation plants successfully reproduce and shed seed, that event may be the start of a population explosion, locally and further afield, provided conditions remain favourable. With a generation cycle of six to eight years, patience is required. Also required is good fortune in the avoidance of hazards. Above-ground parts, especially the inflorescence, may be grazed by deer and rabbits. Below-ground tubers are susceptible to invertebrate attack and may sometimes be rooted out by hungry badgers.

Dehisced Orchid seed falls to the ground mostly close to the mother plant, but the seed dispersal shadow has a very long tail, as the seed is superbly adapted to very long distance dispersal on the wind. It is not unusual to find the occasional Orchid popping up in a restored or created meadow, seed having arrived on the breeze from far off. One individual of an inbreeding species is sufficient to establish a population, while outbreeding species require at least two founder individuals to establish a population.

On the farm where I live we have a home field, a 1.2ha (3-acre) part of which became completely covered in Orchids from seed that must have blown in on the wind. The field was former arable in long-term set-aside, ploughed in 2001, and seeded with Rough Hawkbit (*Leontodon hispidus*) by surface broadcasting and rolling in. This part-field was maintained from then on to the present day by topping (from September to April), during which time Rough Hawkbit has remained dominant. Bee Orchids (*Ophrys apifera*) first appeared in 2007, with numbers peaking in 2009. Other species swiftly followed, making a total of eight species. Population sizes in the part-field were estimated in 2012 from transect counts, and these counts were compared with estimates of the population size for known wild locations for each species at different distances from the home field. The larger populations in the home field were probably second generation.

These observations suggest that the autumn air carries a floating Orchid seed bank and that the home field provided a bioassay of the seeds passing by, with more abundant species from close-by locations being more abundant in the home field. It is unlikely that these species arose from the soil seed bank because the phenomenon was confined to the field with Rough Hawkbit dominant, and absent from adjacent fields (MacIntyre, 2014). Disappointingly, many of the species recorded in 2012 have since declined in numbers, but those of Greater Butterfly-orchid are increasing.

ORCHID SPECIES GROWING IN A PART-FIELD OF *LEONTODON HISPIDUS*, AND DISTANCE TO WILD POPULATIONS OF EACH SPECIES

Estimated population size of each species at increasing distance from the home field

	Distance from home field					
Species	Home field	0–1km	1–2km	2–5km	5–10km	10–20km
Ophrys apifera Bee Orchid	26,000	50–100	1,000–2,000	<10,000	-	-
Dactylorhiza fuchsii Common Spotted Orchid	12,000	100–1,000	2,000–10,000	<100,000	-	-
Anacamptis pyramidalis Pyramidal Orchid	4,000	50–100	1,000–2,000	<10,000	-	-
Dactylorhiza praetermissa Southern Marsh-orchid	25	0	0	0–1	1–10	1,000–10,000
Anacamptis morio Green-winged Orchid	5	0	0	0–1	0–20	10–200
Platanthera chlorantha Greater Butterfly-orchid	3	0	0	0–1	0–10	10–100
Neottia ovata Twayblade Orchid	1	1	1–100	100–1,000	<10,000	-
Gymnadenia conopsia Chalk Fragrant-orchid	1	0	0	0	0-10	-

CHAPTER 16

RESTORATION SPECIES ACCOUNTS

The species accounts that follow briefly describe the 221 proposed restoration species listed in Table 2 (see Appendix). These are species that are proposed for use in habitat restoration and creation in the British Isles. The list includes 195 native species from a native flora of around 1,400 species, 22 archaeophyte species from an archaeophyte flora of around 190 species, and four naturalised neophyte species from a naturalised neophyte flora of around 1,500 species. These are species of the British and Irish flora that are common, widespread and attractive, and are amenable to habitat restoration. Some of these species have a restricted native range within Britain and Ireland, and their use in restoration should be restricted to the native range. Excluded from this list are Schedule 8 protected species, sensitive species, Schedule 9 invasive non-native species, Weeds Act species, other weed species, almost all naturalised neophytes, and all non-naturalised neophytes. Additional descriptive information on restoration species may be found in published botanical accounts and floras (Keble Martin, 1965; Stace, 2022; Stroh et al., 2023).

The accounts that follow give a brief description of the features of each species that are relevant to their use in restoration seeding, and a short account of the seed-production methods for each species, as far as these are known.

1. *Achillea millefolium*: Yarrow (native)

Features: Yarrow is a rhizomatous, grazing-tolerant, perennial flower of pasture and waysides. It is less common in meadows, but widespread and sometimes frequent elsewhere, often forming quite large clonal patches. It is a moderately variable species with white flowers, sometimes pink, with the capitula arranged in umbel-like inflorescences. The fine, silver-coloured seeds germinate readily without dormancy.

Yarrow (*Achillea millifolium*).

Seed of Yarrow.

Seed production: Row-crop establishment is by precision drilling at a field scale in the autumn or spring. The crop grows well, flowering in the second full year after drilling. Harvest is by combine harvester with the concave half closed and aspiration minimal. The yield of material will be huge, and cleaning will be needed to remove floret parts, bracts and fine leaf fragments. The amount of seed obtained is usually good.

2. *Achillea ptarmica*: Sneezewort (native)

Features: Sneezewort is widespread, but scattered, uncommon and declining. It is a perennial of wet grasslands and wet meadows on neutral to acidic soils, and is more frequent in the north.

Seed production: Sneezewort may be grown in short rows and the seed heads harvested by hand or combine harvester.

3. *Adonis annua*: Pheasant's-Eye (archaeophyte)

Features: Pheasant's-eye is a striking, cushion-forming annual of cultivation and disturbed soil, now quite rare in the wild as a consequence of chemical farming. The seeds appear to have complex dormancy-breaking requirements and do not germinate readily.

Seed production: Pheasant's-eye is a species that was formerly more widespread, and its reintroduction would be good; however, its cultural requirements are not fully known and it remains a difficult species to grow in cultivation.

4. *Aethusa cynapium*: Fool's Parsley (native)

Features: Fool's Parsley is a common, white-flowered spring-annual Umbellifer of cultivated ground.

Seed production: Fool's Parsley will often spring from cultivated ground on a nursery or farm, and seed is easily collected by hand in moderate quantities.

Fool's Parsley (*Aethusa cynapium*).

5. *Agrimonia eupatoria*: Agrimony (native)

Features: Agrimony is a long-lived perennial with yellow, lemon-scented flowers, and is frequent along waysides and hedgerows, except where cut in spring or summer. The large conical seeds sport a ring of hooked appendages, and the seeds are borne on tall, unbranched spikes waiting to hitch a ride. To brush too close to a plant can be very annoying for livestock and pets.

Seed production: Germination of Agrimony seed can be erratic, and establishing row crops from drilled seed is not easy. However, mature seed can often be plucked by hand from waysides and hedgerows, and foraging in this way is probably better than attempting to grow Agrimony in rows.

Agrimony (*Agrimonia eupatoria*).

Seed of Agrimony.

6. *Agrostemma githago*: Corn Cockle (archaeophyte)

Features: Corn Cockle is a tall, large-seeded, winter-hardy annual of cultivated ground, now extremely rare in the wild, but frequently sown.

Seed production: Sow Corn Cockle in late summer or early autumn. The seed will germinate to produce seedlings that over-winter and go on to produce flowering plants in the following year. Harvesting is by combine harvester, and this can produce enormous quantities of seed.

Corn Cockle (*Agrostemma githago*).

Seed of Corn Cockle.

7. *Agrostis capillaris*: Common Bent (native)

Features: Common Bent is a rhizomatous, fine-leaved perennial grass, common, widespread and abundant in the wild. It is not a competitive or tall-growing grass. At flowering, the very fine-textured inflorescences are borne above the leaves, and en masse in a breeze, appear like passing mist.

Seed production: Common Bent is a proscribed species subject to the Seed Marketing Regulations 2011. Wild-type native stock is readily grown as a row crop at a field scale, and is harvested in the normal way by combine harvester.

8. *Ajuga reptans*: Bugle (native)

Features: Bugle is a widespread and declining stoloniferous spring-flowering perennial of woodland and hedgerow. It is very attractive in flower, and can form large clonal patches of characteristically glossy, dark-green leaves. Like so many species capable of vegetative spread, it is modest at producing seed. In

Bugle (*Ajuga reptans*).

flower, Bugle is similar to, and may be confused with Selfheal (*Prunella vulgaris*).

Seed production: Bugle is best propagated vegetatively from rooted stolons, rather than from seed.

9. *Alliaria petiolata*: Garlic Mustard (native)

Features: Only faintly smelling of garlic, and not tasting of mustard, Garlic Mustard is a frequently encountered and attractive spring-flowering biennial of hedgerows. It is frequented by the Orange Tip butterfly (*Anthocharis cardamines*) and is host to the caterpillars of the Orange Tip. Garlic Mustard can be invasive of woodland ground flora outside its native range (Bialic-Murphy et al., 2019).

Seed production: Garlic Mustard is a biennial and grows well in cultivation. From an autumn or spring drilling it will flower in the first full year after sowing, and produce clusters of narrow seed-containing pods. These may be picked by hand or harvested by combine harvester. Garlic Mustard is a high-yielding species.

Garlic Mustard (*Alliaria petiolata*).

Seed of Garlic Mustard.

10. *Allium ursinum:* Ransoms (native)

Features: Ransoms (Wild Garlic) has leaves that are intensely garlic flavoured, and these are edible in moderation. It is nice uncooked fresh, diced and mixed into mashed

Ransoms (*Allium ursinum*).

Seed of Ransoms.

potato. It is a spring-flowering, tuber-forming perennial, providing abundant early pollen and nectar; it is common in woodland, and often dominates the ground flora on more alkaline soils. The black seeds require vernalisation to break dormancy, and are best sown in the autumn.

Seed production: Seed may be collected by hand from woodland (with permission) where Wild Garlic is the dominant ground cover. Seed set will vary from year to year, and is best when early spring weather is sufficiently warm to awaken hibernating pollinators.

11. *Alopecurus geniculatus*: Marsh Foxtail (native)

Features: Marsh Foxtail is a perennial grass of wet meadows, marshes and ditches. It is frequent and common throughout the British Isles.

12. *Alopecurus pratensis*: Meadow Foxtail (native)

Features: Meadow Foxtail is a meadow perennial and is the first of the common meadow grasses to flower, followed by Sweet Vernal-Grass. It is a frequent constituent of lowland meadows, but rather less frequent on dry soils. The flower heads are pale buff in colour, soft to the touch, and similar in appearance to Timothy, which is very much later flowering.

Seed production: Row crops of Meadow Foxtail may be established by precision drilling, and will grow well in cultivation. The light seed can be combine harvested with minimal aspiration.

Meadow Foxtail (*Alopecurus pratensis*).

Seed of Meadow Foxtail.

13. *Anacamptis morio*: Green-winged Orchid (native)

Features: Green-winged Orchid (former Latin name: *Orchis morio*) was the iconic species of the traditional, species-rich lowland hay meadow (Stroh, 2015a), but has suffered a

Green-winged Orchid (*Anacamptis morio*).

catastrophic decline in the wild as a consequence of the switch from hay to silage and the adoption of chemical farming, and is rarely seen these days. It is a wintergreen, early flowering Orchid and is able to flower and release seed before a late June/early July hay cut, but not before a May silage cut. The leaves are unspotted and the plant is polycarpic, and produces flowers ranging in shade from the typical purple, to pink, to white. All flowers display a defining pair of parallel-veined upper tepals. Recovery of wild populations of Green-winged Orchid would be a sure and pleasing sign that nature restoration is working, but we may have to wait decades to see it happen, if at all. Like all Orchids, Green-winged Orchid will propagate in the wild from sown seed, but only in the presence of the Orchid-specific mycorrhizal fungal species that it requires (Ercole et al., 2014), and only if the ground is subject to long-term beneficial management.

Seed production: Seed-bearing inflorescences may be collected from plants in cultivation, but not from the wild.

14. *Anacamptis pyramidalis*: Pyramidal Orchid (native)

Features: Pyramidal Orchid is a moderately common and widespread species in the wild, more frequent in the south than further north, but declining in most places. The leaves are unspotted and narrow lanceolate, and the plant is a polycarpic perennial, flowering in mid-summer just after the Common Spotted Orchid. Seed dehiscence is around mid-August, well after normal cutting dates for meadows. Pyramidal Orchid is often present in meadows, sometimes in quantity. These plants will not form seed if cut at flowering, so it may be good to delay the hay cut till after mid-August, or to leave patches uncut, to allow seed maturation and release.

Seed production: Seeding inflorescences of Pyramidal Orchid may be collected from plants in cultivation, or from plants in meadow and pasture that would otherwise be cut or grazed, but never from the wild. The seedpods of Pyramidal Orchid are small compared to those of Common Spotted Orchid, and seed yield is also small.

Pyramidal Orchid (*Anacamptis pyramidalis*).

Wood Anemone (*Anemone nemorosa*).

15. *Anemone nemorosa*: Wood Anemone (native)

Features and seeds: Wood Anemone is a very attractive, widespread and frequent spring woodland perennial. It is one of very few species native to the British Isles that has recalcitrant seeds. Recalcitrant means that the seeds at maturity are intolerant of drying and must be stored fully imbibed, or sown directly without drying.

16. *Angelica sylvestris*: Wild Angelica (native)

Features: Wild Angelica is a tall, biennial Umbellifer of wet grasslands, marshes, ditches and the waterside.

Seed production: Wild Angelica can be grown as a row crop; small plots can be harvested by hand, and larger plots by combine harvester.

Wild Angelica (*Angelica sylvestris*) with feasting pollinators.

Seed of Wild Angelica.

17. *Anthoxanthum odoratum*: Sweet Vernal-grass (native)

Features: Sweet Vernal-grass is a widespread and abundant native annual grass of unimproved (agriculturally) meadows across the British Isles. It is early heading, very palatable to livestock, and comes into anthesis soon after Meadow Foxtail; it is sweet smelling, and responsible for the sweet smell of new-mown wild hay.

Sweet Vernal-grass (*Anthoxanthum odoratum*) at anthesis.

Seed of Sweet Vernal-grass with appendages removed.

Seed production: Sweet Vernal-grass is early to shed seed, dropping seed in early June, and may not be properly represented in brush-harvested seed samples taken in early July. The seed at shedding resembles a flea. It is small, brown, a little fluffy, and has an angled awn that can flick on changes in humidity, so the seed jumps. Sweet Vernal-grass can be precision drilled and grown as a row crop. The seed may be harvested by combine harvester. The yield is weather and timing critical, but can be good.

18. *Anthriscus sylvestris*: Cow Parsley (native)

Features: Cow Parsley is the first to flower in a succession of white-flowered Umbellifer species that line hedges and waysides from spring onwards. Flowering of Cow Parsley is soon followed by Hogweed, then Rough Chervil, and finally Upright Hedge Parsley. These are all palatable species, loved by equines, cattle and sheep, and are usually grazed out and largely absent from the in-field side of a hedge. The name 'Cow Parsley' is a reference to the species palatability.

Seed of Cow Parsley.

Cow Parsley (*Anthriscus sylvestris*).

Seed production: Cow Parsley is a very common hedgerow plant, and seeds may be collected by hand in early summer when the clusters turn from green to black.

19. *Anthyllis vulneraria*: Kidney Vetch (native)

Features: Kidney Vetch is a yellow-flowered, wintergreen biennial Vetch, frequent on the coast and sometimes on

Kidney Vetch (*Anthyllis vulneraria*) on Chesil Bank.

Seed of Kidney Vetch, green at one end, after removal from the black pod.

light soils inland. It is a variable species, with some local ecotypes having red-edged flowers. The kidney-shaped seeds are held in a one-seeded black pod retained within a persistent calyx. The native wild-type ecotype with yellow flowers is widely used in restoration seeding.

Seed production: Kidney Vetch grows well in cultivation, but seed harvest and seed extraction is challenging and requires dedicated machinery.

20. *Armeria maritima*: Thrift (native)

Features: Thrift is a coastal perennial halophyte, and occasionally, in recent years, it has been spotted on inland salted road verges. It is an attractive species forming tight cushions of fine leaves from which inflorescences arise with clusters of pink flowers on an unbranched peduncle. The seed, one per floret, is enclosed in a dart-shaped persistent calyx that acts as a buoyancy aid, assisting dispersal by water and wind. A rare, long-peduncled inland form (*A. maritima ssp elongata*) is native to a few localities in Lincolnshire.

Seed production: Thrift may be grown in small plots using horticultural methods. Mature plants at seeding produce clusters of light-coloured papery seeds that can be collected by hand. Seed yield is low.

21. *Arrhenatherum elatius*: False Oat-grass (native)

Features: False Oat-grass is the dominant grass of waysides and rough ground in the British Isles. Newly formed road verges are often colonised first by annual species, then by pioneer biennials including ragwort, docks and thistles, then by False Oat-grass. With occasional scrub control, vegetation dominated by False Oat-grass quickly becomes

Thrift (*Armeria maritima*) on Chesil Bank.

False Oat-grass (*Arrhenatherum elatius*).

Awned seed of False Oat-grass.

the climax vegetation of many roadside verges on rich to moderately fertile soils. False Oat-grass is a tufted perennial that responds positively to a lack of grazing or mowing. Forms of the species exist that have bulb-like swellings at the stem bases, and these forms are known as Onion Couch.

Seed production: False Oat-grass is a proscribed species subject to the Seed Marketing Regulations 2011. Its use in restoration seeding is minimal, but when required it may be collected by hand from waysides.

22. *Arum maculatum*: Lords-and-ladies (native)

Features: Lords-and-ladies, known also by many other different colloquial names, is a rhizomatous perennial frequent in woodland and along hedge lines. The spotted leaves, the spathe and the spadix are characteristic. Italian Lords-and-ladies (*Arum italicum*) is also native to the British Isles, and is distinguished by being less spotted and having a paler spadix.

Lords-and-ladies (*Arum maculatum*).

Seed of Lords-and-ladies.

Seed production: Lords-and-Ladies may be propagated from the red berries that can be collected in late summer.

23. *Asplenium scolopendrium:* Hart's-tongue (native)

Features: Hart's-tongue Fern (former Latin name: *Phyllitis scolopendrium*) is a frequently encountered fern of woodland,

Hart's-tongue Fern (*Asplenium scolopendrium*) with leaves unfurling crozier-like.

hedgerow and wayside. The strap-like undivided fronds are characteristic, and it propagates by dust-like spores rather than seeds, as do all ferns. Spores may be collected from the linear sori on the underside of fronds. Non-seed-bearing plants, ferns, horsetails, mosses, liverworts and so on are all very important constituents of terrestrial ecosystems, but as they reproduce by spores rather than seeds, their planned incorporation into restoration schemes may sometimes be difficult. Generally it is good to encourage the natural arrival of lower plants by habitat provision.

24. *Avenella flexuosa*: Wavy Hair-grass (native)

Features: Wavy Hair-grass (former Latin name: *Deschampsia flexuosa*) is a tufted perennial widespread on acid heaths and moors throughout the British Isles; it is often the dominant grass in these situations.

Seed production: Wavy Hair-grass is a key species for use in the restoration of moor and heath; seed can sometimes be machine harvested in quantity from wild stands.

25. *Avenula pubescens*: Downy Oat-grass (native)

Features: Downy Oat-grass (former Latin name: *Helictotrichon pubescens*) is a tufted perennial of calcareous soils over limestone or chalk throughout the British Isles. Rarely dominant, it is nevertheless an important constituent of wild calcareous grasslands and is amenable for use in the restoration of these grasslands as it establishes readily from sown seed.

Downy Oat-grass (*Avenula pubescens*) at anthesis.

26. *Barbarea vulgaris*: Winter-cress (native)

Features: Winter-cress is an attractive, yellow-flowered biennial, sometimes winter annual, Crucifer of rough ground, disturbed soils and arable cultivation. Unlike the similar *Sinapis arvensis* (Charlock), Winter-cress is not considered a weed of cultivation.

Seed production: Winter-cress establishes readily from late summer-drilled seed, at first producing a rosette of dark-green shiny leaves that are hardy and over-winter in our climate; it then proceeds to flower in the early summer of the following year. The seed may be harvested by combine harvester, and seed yields are good.

Winter-cress (*Barbarea vulgaris*).

27. *Bellis perennis*: Daisy (native)

Features: Familiar to all, the common wild Daisy is frequent and abundant in meadows, mown amenity grassland, and grazed pastures across Britain and Ireland. It is a low-growing, rosette-forming perennial, and has a long indeterminate flowering period. The interval between flowering and seed shed is very short, hence the species' success in short mown and grazed swards.

Daisy (*Bellis perennis*).

Seed of Betony.

Seed production: Daisy has fine seed that it is not easy to collect, even from patches with abundant flowers.

28. *Betonica officinalis*: Betony (native)

Features: Betony (former Latin name: *Stachys officinalis*) is a long-lived perennial of hedge banks, heaths and grasslands. It is a declining species that is widespread and locally frequent in England and Wales, but largely absent from Scotland and Ireland. The flowers are typically mauve in colour, sometimes mixed with white and intermediate pink-flowered forms.

Colour variants of Betony (*Betonica officinalis*).

Seed production: Row crops of Betony may be established by precision drilling in the autumn; plants may take two to three years to establish and reach reproductive maturity. Mature plants will yield good quantities of seed for several years, and Betony seed is easy to harvest by combine harvester.

29. *Blackstonia perfoliata*: Yellow-wort (native)

Features: Yellow-wort is an uncommon annual of calcareous grassland, dunes and exposed chalk and limestone.

Yellow-wort (*Blackstonia perfoliata*) amongst Harebell (*Campanula rotundifolia*) and Common Centaury (*Centaurium erythraea*).

The distinctive leaves are glaucous and fused in pairs around the stem. The seed is dust-like and black.

Seed production: Growing Yellow-wort in cultivation is difficult, requiring horticultural methods and attention to detail.

30. *Brachypodium sylvaticum*: False Brome (native)

Features: False Brome is common throughout the British Isles, growing along hedgerows, field margins and waysides. It is a good species for restoration of these habitats.

Seed production: False Brome is easily grown in cultivation at a field scale and can be combine harvested, producing large quantities of seed.

False Brome (*Brachypodium sylvaticum*).

31. *Briza media*: Quaking Grass (native)

Features: Quaking Grass is probably the prettiest grass of the meadow, with numerous purse-shaped hanging florets that flutter in a breeze. It is a long-lived tufted perennial, once widespread and common in unimproved neutral and calcareous grasslands across the British Isles.

Seed production: Quaking Grass may be precision drilled in rows. From drilling to reproductive maturity is around three years. Established crops of Quaking Grass will continue to yield seed for many subsequent years. The seed may be combine harvested with the concave set tight and aspiration minimal.

Quaking Grass (*Briza media*) showing inflorescence unfurling.

Seed of Quaking Grass.

32. *Bromopsis erecta*: Upright Brome (native)

Features: Upright Brome is frequent and abundant, sometimes dominant, in wild calcareous grasslands across England, but only occasional in Scotland, Wales and Ireland. Like Quaking Grass, Upright Brome is a long-lived tufted perennial, not reaching reproductive maturity until five to six years after restoration seeding. It is a very good species to use in the restoration and creation of calcareous meadows because it creates an open structure at different heights within the sward, allowing for the co-existence of different community component species and high diversity. In pasture that is extensively grazed and not topped Upright Brome can create a slightly less species-diverse tussock structure.

Seed production: In some wild calcareous grasslands, Upright Brome can be dominant, providing an easy opportunity to harvest see of reasonable purity by brush harvester or combine harvester, direct from the wild, where this is allowed.

Upright Brome (*Bromopsis erecta*) at anthesis.

Upright Brome seed.

33. *Calluna vulgaris*: Heather (native)

Features: Heather is a key and common species of heaths and moors on peaty or sandy soil, and is

Heather (*Calluna vulgaris*) on Gower Peninsula.

sometimes dominant. It is a perennial low-growing shrub, widely dispersed across suitable habitat in Ireland, Scotland, Wales and England. The fine seed has good longevity in the soil, and plants will often appear on disturbed soil where disturbance exposes the soil seed bank.

Seed production: Heather seed may be collected from existing wild stands using a brush harvester, where permitted. The harvested material will contain fine seeds suspended in a mixture of florets and leaves, and this mixture may be sown directly, before or after drying. Extraction of pure Heather seed from a brush-harvested mixture can be challenging.

34. *Caltha palustris*: Marsh Marigold (native)

Features: Marsh Marigold is an attractive and occasional perennial of wet meadows, ditches, ponds and wet woods, frequent across Ireland, Scotland, Wales and England. Polyploid forms of the species are present in the wild, and they may be recognised by noticeable differences in leaf and flower size.

Seed production: Marsh Marigold is suitable for modest scale horticultural production of seed. Raised plants grow well if planted into moist ground, and seed may be harvested by hand as the pods split lengthways.

Marsh Marigold (*Caltha palustris*) at Arlington Row, Bibury.

Seed of Marsh Marigold.

35. *Campanula glomerata*: Clustered Bellflower (native)

Features: Clustered Bellflower is a short, erect perennial largely confined to unimproved calcareous grasslands in England. It is declining and uncommon.

Clustered Bellflower (*Campanula glomerata*).

Seed production: The fine seeds of Clustered Bellflower germinate without dormancy, and plants may be grown from seed using horticultural methods. Seed harvesting is normally by hand, gathering seeding clusters and drying on sheets to release the seed.

36. *Campanula rotundifolia*: Harebell (native)

Species features: Harebell is heterophyllus, having basal leaves that are round in shape, and stem leaves that are elongate lanceolate. It is a rhizomatous perennial, a low-growing plant of thin stature, and it can be found growing on a wide range of different soil types across Scotland, Wales, England and NW Ireland. In central and eastern areas it is represented mostly by tetraploid cytotypes

Harebell (*Campanula rotundifolia*).

Seed of Harebell.

(2n=4x=68) and in western areas by hexaploid cytotypes (2n=6x=102). Restoration seeding should ensure, if possible, that the home cytotype is used. Harebell survives well on poor, droughty soils.

Seed production: In cultivation, Harebell is not competitive, requiring a lot of attention to achieve good-sized plants. Seed harvesting is by hand, cutting whole plants at yellow pod stage and placing the material on a sheet to dry. Seed yield is always very low, but the seed, when obtained, is pleasing to handle and will germinate without dormancy. *The Biological Flora for Harebell* has been published (Stevens et al., 2012).

37. *Campanula trachelium:* Nettle-leaved Bellflower (native)

Features: Nettle-leaved Bellflower is an upright perennial growing to 1m (3ft) with Nettle-like leaves – it may be mistaken for a Stinging Nettle when not in flower. The main habitat is woodland and hedgerows on calcareous soils, and it is mostly confined to central England.

Seed production: Row crops of Nettle-leaved Bellflower may be established by precision drilling in the field. It grows well, but is very susceptible to rust (*Puccinia campanulae*) in our climate. Surviving crops may be combine harvested, yielding well for a *Campanula* species. The seed germinates without dormancy and appears to be very mobile in the surrounding environment, although the seeds seem to have no obvious adaptations for dispersal.

Nettle-leaved Bellflower (*Campanula trachelium*).

Seed of Nettle-leaved Bellflower.

38. *Capsella bursa-pastoris*: Shepherd's Purse (archaeophyte)

Features: Shepherd's Purse may be considered a garden weed, unwelcome by some, but in the wild it is a familiar and beneficial annual colonist of disturbed and cultivated soils, being one of the most common species on our planet. It is not invasive and has a beneficial early role in the succession from bare soil to mature herbaceous vegetation.

Seed production: The purse-shaped capsules of Shepherd's Purse are easily collected from plants growing naturally on disturbed ground; on drying they release fine yellow seeds.

Shepherd's Purse (*Capsella bursa-pastoris*).

39. *Cardamine pratensis*: Lady's Smock (native)

Features: Lady's Smock is a ubiquitous perennial herb of wet meadows, grassy places and woodland, present throughout Ireland, Scotland, Wales and England. Plants are very variable, and flower colour ranges from white to pink.

Lady's Smock (*Cardamine pratensis*).

Seed production: Some forms of the species produce a modest amount of seed, others none, relying more on readily detached leaflets that root into the ground where they fall. Lady's Smock is an important constituent of wet grasslands, but is not amenable to producing seeds in quantity in cultivation.

40. *Carduus nutans*: Musk Thistle (native)

Features: Musk Thistle is a handsome biennial, sometimes perennial, of waysides and grasslands on calcareous soils, confined mostly to England and Wales.

Seed production: Musk Thistle grows well in cultivation and produces a moderate amount of seed when combine harvested.

41. *Carex divulsa*: Grey Sedge (native)

Features: Grey Sedge is a densely tufted deciduous perennial of hedgerows, waysides and grassy places. Currently frequent in southern Ireland and central and southern England, it is expanding its range northwards, possibly in response to climate change. Grey Sedge is a robust

Musk Thistle (*Carduus nutans*) with characteristic nodding head.

perennial creating a good tussock structure in unmanaged grassland.

Seed production: Grey Sedge may be precision drilled in rows and is an easy species to grow in cultivation. It may be combine harvested to produce large quantities of seed for use in restoration projects.

Grey Sedge (*Carex divulsa*).

Seed of Grey Sedge.

42. *Carex echinata:* Star Sedge (native)

Features: Star Sedge is a densely tufted, deciduous perennial of rough grasslands, bogs and marshes, widespread in Ireland, Scotland, Wales, and a little less so in England.

Seed production: Star Sedge is easy to grow in cultivation and may be combine harvested to produce very good quantities of seed.

Star Sedge (*Carex echinata*).

Seed of Star Sedge.

43. *Carex flacca*: Glaucous Sedge (native)

Features: Glaucous Sedge is a widespread, rhizomatous, wintergreen perennial of calcareous and neutral grasslands. It is low growing, and the leaves are distinctly glaucous.

Glaucous Sedge (*Carex flacca*).

Seed of Glaucous Sedge.

Seed production: Glaucous Sedge is slow to establish and slow growing in cultivation, but very long lived, often forming large clonal patches. It is an early flowering sedge, and the seeds are retained on the inflorescence for some time, allowing for easy harvest by combine harvester, with the drum and aspiration set at moderate. Seed yield is modest.

44. *Carex pendula*: Pendulous Sedge (native)

Features: Pendulous Sedge is a wintergreen, large tufted perennial Sedge of wet woods, ditches and streamsides, distributed throughout the British Isles and frequent in the south. It can form dense tussocks under favourable conditions, producing long drooping leaves and pendulous flowering stems to 1.8m (5ft).

Seed production: Seed is easily hand collected from wild stands or from plants in cultivation, although yield is modest.

Juvenile plant of Pendulous Sedge (*Carex pendula*).

Seed of Pendulous Sedge.

45. *Carex sylvatica*: Wood Sedge (native)

Features: Wood Sedge is an early flowering, wintergreen, perennial Sedge forming rather handsome tussocks in woodland.

Seed production: Wood Sedge grows poorly in cultivation on open ground, and seed production, even in its natural setting, is quite modest.

Wood Sedge (*Carex sylvatica*).

Seed of Wood Sedge.

46. *Carlina vulgaris*: Carline Thistle (native)

Features: Carline Thistle is a declining biennial Thistle of small stature found growing in open situations on infertile calcareous soils, on fixed dunes and on base-rich rocky ground. It is largely confined to coastal areas in Scotland, but distributed throughout most of the rest of the British Isles, if rather sparingly.

Seed production: Carline Thistle is a challenging species, and guidance on growing and seed harvest awaits advances in this area.

Carline Thistle (*Carlina vulgaris*).

Blue- and white-flowered Cornflower (*Centaurea cyanus*).

47. *Centaurea cyanus*: Cornflower (archaeophyte)

Features: Cornflower was present in Britain from the Iron Age onwards as a winter annual naturalised in cornfields, but is now very rare as it is easily controlled by modern herbicides.

Seed production: Cornflower is easily grown in cultivation at a field scale, precision drilling in rows in late summer before temperatures fall. The seed has moderate to low longevity in the soil, and requires exposure to light and an optimum temperature of 10–150C for germination (Guillemin et al., 2017). A combine harvester can readily harvest the seed, with the combine drum and aspiration set at moderate. Seed yield is very good. It is an excellent species for growing in annual mixtures, and is superb at providing pollinator services.

Seed of Cornflower.

48. *Centaurea nigra*: Common Knapweed (native)

Features: Common Knapweed is an attractive and widespread robust perennial of meadows, pastures, downland, floodplain meadows, waysides and uplands, on substrates varying in extremes of wetness, and in pH from moderately acid to quite alkaline. It is a highly variable species found throughout Britain and Ireland as a wide range of distinct ecotypes in both rayless and rayed forms. It is a generalist species of our islands and much used in restoration seeding, although it can be invasive outside its native range. The seed has a reduced pappus of short bristles, an elaiosome at the point of attachment (hilum) that may aid dispersal by ants (Skilbeck, 2023), and no dormancy requirements.

Rayless form of Common Knapweed (*Centaurea nigra*).

Seed of Common Knapweed.

Seed production: Row crops of Common Knapweed may be precision drilled at a field scale in the autumn or spring. Crop establishment is moderately quick, with first flowering in the second full year after drilling. Flowering crops attract huge numbers of pollinators, and after pollination the crop will often attract flocks of seed-robbing finches, usually a sign to get the combine out. The seed is held tight within capitula and cannot be readily extracted by hand. This is not a problem for a combine harvester with the drum set closed and aspiration at medium. However, not all seed is extracted at the first combine harvest, and additional yield can be obtained by picking up the crop again after a few days drying in the swathe. Overall the seed yield is very good – until the Spotted Knapweed Seed Head Moth (*Metzneria paucipunctella*) finds the crop. The larvae of the moth are a serious pest of seed crops of Common Knapweed in cultivation, usually moving in after a few years of cropping, and the only solution to this invasion is to relocate cropping to another location.

49. *Centaurea scabiosa*: Greater Knapweed (native)

Features: Greater Knapweed is a robust perennial of dry, calcareous soils; it is a widespread and frequent species of meadows, pastures, downland and waysides on chalk and limestone. It is a variable species and differs from Common Knapweed in having more dissected leaves, involucre bracts that are silvery, and larger flowers that are invariably rayed. The seed is large, glossy brown and has a conspicuous pappus of long bristles.

Seed production: Greater Knapweed is an important restoration species and may be easily cultivated for seed, but only on high pH ground. Harvest is by combine harvester with the drum set closed and the aspiration up. Caution is needed to avoid contact with the irritant sterile hairs of the capitulum, of which there are masses. Seed yield is very good. Greater Knapweed is susceptible to Spotted Knapweed Seed Head Moth larvae, but a little less so than Common Knapweed.

Greater Knapweed (*Centaurea scabiosa*).

Seed of Greater Knapweed with pappus partially removed.

50. *Centaurium erythraea*: Common Centaury (native)

Features: Common Centaury is an infrequent biennial herb of short open grassland on mildly acidic to calcareous, nutrient-poor, free-draining soils. It is a declining species that may be found growing in nutrient-poor grasslands, sand dunes, open woodland rides and on disturbed ground.

Common Centaury (*Centaurium erythraea*).

51. *Cerastium fontanum*: Common Mouse-ear (native)

Features: Common Mouse-ear is a common constituent of grasslands throughout the British Isles, but is rather diminutive and often overlooked. It is a low-growing, matt-forming perennial with small white flowers.

Seed production: Common Mouse-ear is not amenable to cultivation for seed production, but the small yellow seeds will often be present in brush-harvested mixtures. The seeds may be separated with careful cleaning and sorting.

Common Mouse-ear (*Cerastium fontanum*).

52. *Chaenorhinum minus*: Small Toadflax (archaeophyte)

Features: Small Toadflax is a spring annual of well-drained, calcareous soils. Formerly a common weed of arable cultivation, it is now a rare arable weed. It survives elsewhere on disturbed ground, tracks and wasteland.

Seed production: Small Toadflax is not easily cultivated, but with a change of management on a restoration site it is a species to look out for as it may, with luck, germinate from the soil seed bank following disturbance.

Small Toadflax (*Chaenorhinum minus*).

53. *Chaerophyllum temullum*: Rough Chervil (native)

Features: Rough Chervil is a moderately tall white-flowered biennial Umbellifer, similar looking to Cow Parsley (*Anthriscus sylvestris*), which it follows after in the spring flowering sequence of waysides and hedgerows. It is tolerant of semi-shade and is frequent along hedgerows, waysides and roadsides. Rough Chervil is considered to be a neophyte in Ireland.

Rough Chervil (*Chaerophyllum temullum*).

Seed of Rough Chervil.

Seed production: The seed requires vernalisation and germinates well in the spring from an autumn drilling. Under good conditions, a seed crop will flower and mature in the second full year after drilling and may be combine harvested, yielding very well.

54. *Circaea lutetiana*: Enchanter's Nightshade (native)

Features: Enchanter's Nightshade is a rhizomatous perennial herb of moist, shady habitats in woodland, scrub, hedgerows and streamside. Flowers are borne on an erect, sometimes branched inflorescence, and the easily detached fruit is a one- or two-seeded achene clothed in small hooks, providing a means of dispersal.

Seed production: Enchanter's Nightshade is a very common species, and seed may often be collected from hedgerows by hand. In the garden it can be a pest, proliferating by persistent rhizome fragments.

Enchanter's Nightshade (*Circaea lutetiana*).

55. *Cirsium acaule*: Dwarf Thistle (native)

Features: Dwarf Thistle is a rosette-forming, ground-hugging perennial of short and grazed calcareous grassland and pasture, with flowers borne on very short pedicels in the centre of rosettes. It is highly tolerant of grazing, but is not competitive in erect vegetation, and is not a frequent meadow species. Dwarf Thistle is found growing in the southern third of the British Isles from Yorkshire south, and may be expected to move north with global warming. The seed has a fluffy pappus and is wind dispersed.

Seed production: The ground-hugging habit precludes mechanical seed harvesting, although hand collection of small quantities of seed is possible on a still day.

Dwarf Thistle (*Cirsium acaule*).

56. *Cirsium eriophorum*: Woolly Thistle (native)

Features: Woolly Thistle is the largest and most majestic of our native thistles. It is a monocarpic perennial/biennial, occasional on dry, often ungrazed grasslands, pasture,

Woolly Thistle (*Cirsium eriophorum*).

open scrub and disturbed ground on calcareous soils in the southern half of the British Isles.

Seed production: Woolly Thistle may be grown in cultivation on calcareous soils. The large capitula close after pollination, and protected by leather gloves, seed heads may be removed individually with secateurs. On drying these may partially open, and when fully dry the seed can be extracted by passing the material through a static combine, taking all necessary precautions. Seed yield can be very good.

57. *Cirsium palustre*: Marsh Thistle (native)

Features: Marsh Thistle is a widespread monocarpic perennial/biennial of damp places, found throughout Britain and Ireland. It can be found growing in mires, fens, marshes, damp grassland, rush pastures, wet woodland edges, montane springs and flushes, and tall-herb vegetation. It can be variable in flower and leaf pigmentation, and although thistle-like in appearance, it is not an invasive weed of agricultural grasslands, providing significant ecosystem services for the modest space it occupies.

Marsh Thistle (*Cirsium palustre*).

Seed production: Marsh Thistle seed heads may be hand harvested from plants growing on farmland where permission to collect has been given.

58. *Clematis vitalba*: Traveller's Joy (native)

Features: Traveller's Joy is a late summer flowering, liana-like long-lived perennial of calcareous soils, native south of The Wash. It is moderately invasive and competitive, and can clamber over and cover trees, scrub, hedges, hedge

Traveller's Joy (*Clematis vitalba*).

banks, walls and sand dunes. Traveller's Joy is palatable and liked by livestock, so is often absent from the field side of a hedge while abundant on the track side. It is a pest to hedge trim, and will wrap around the flail head.

Seed production: The seed has a feathered appendage, and despite an apparent abundance of seed in early winter there is no easy way to collect the seed other than by hand.

59. *Clinopodium vulgare*: Wild Basil (native)

Features: Wild Basil is a declining, aromatic, short-lived perennial of calcareous soils, widespread and occasional across Britain and Ireland. It may be found growing in grassland, hedgerows and scrub, and colonises disturbed ground.

Seed production: Wild Basil is amenable to cultivation, and row crops can be easily established by precision drilling. Crops grow well, and mature to produce seed that is retained on the inflorescence for a reasonable period of time. Seed may be combine harvested with the drum half closed and the aspiration minimal. It is not high yielding, but is a pleasing crop to grow because of the huge number of invertebrates that it attracts.

60. *Conopodium majus*: Pignut (native)

Features: Pignut is a spring-flowering deciduous perennial Umbellifer of meadows, pastures, waysides, hedgerows, and sometimes woodland. It is found throughout most of Britain and Ireland, and is particularly characteristic of some northern meadows. The leaves are very finely divided and the flowers are delicate. After seeding, plants die back to a 'nut' just below the soil surface, much loved by badgers in winter. They dig in patches at night and leave the ground very disturbed.

Seed production: The seed of Pignut may often be collected from fields where the species is abundant, by brush harvesting or hand picking; however seed yield is usually very modest.

Wild Basil (*Clinopodium vulgare*).

Pignut (*Conopodium majus*).

61. *Crambe maritima*: Seakale (native)

Features: Seakale is a fleshy-leaved, long-lived perennial halophyte largely restricted to coastal shingle and nearby locations. It is only occasional in the wild, and wild seed, or plants, should not be collected. Seakale is easily grown in the garden and may be 'forced', to produce very tasty edible leaves in spring, when cooked.

Seed production: In cultivation Seakale can produce good quantities of seed, each seed borne individually on a short pedicel. The seeds are encased in a buoyant pericarp, aiding dispersal. The pericarp does not hinder germination and does not need to be removed. *The Biological Flora of Seakale* has been published (Sanyal and Decocq, 2015).

Seakale (*Crambe maritima*).

62. *Crithmum maritimum*: Rock Samphire (native)

Features: Rock Samphire is a fleshy, summer-flowering perennial halophyte with corky fruits. It is frequent, and sometimes abundant, on spray-drenched sea-cliffs, coastal rocks and stabilised shingle in the West, South and East of Britain and Ireland.

Seed production: Rock Samphire produces abundant corky seed, and seed collection is easy where access is good, and where collection is permitted.

63. *Cruciata laevipes*: Crosswort (native)

Features: Crosswort is a sweet-scented, mat-forming perennial of ungrazed grassland, open scrub, woodland edge, rides and waysides, frequently found in Devon, the Cotswolds, the North and South Downs, and the Isle of Wight. Flowering shoots are unbranched, with small, simple

Crosswort (*Cruciata laevipes*).

Seed of Crosswort.

leaves arranged in whorls of four, hence the name Crosswort. Yellow flowers, each with four petals, are borne individually on short pedicels in the axils of leaves, and after pollination the pedicel curves down to hide the developing seed beneath the whorl of small leaves, which also curve to hide the seeds. Seeds are retained in this position and as they mature they turn from green to brown in colour.

Seed production: In cultivation, Crosswort can be precision drilled in rows in the autumn and will germinate in the spring after vernalisation in the soil. It will then establish easily and go on to produce large yields of seed for several years. The seed may be combine harvested with a header fitted with crop lifters. Crosswort is an all-round good restoration species and may be used within its natural native range.

64. *Cynoglossum officinale*: Hound's-tongue (native)

Features: Hound's-tongue is a tall, large-leaved, summer-flowering biennial of disturbed ground on dry soils. It is found on coastal sand dunes and shingle, open grassland, woodland margins and clearings, field edges, cleared land and gravelly waste. It has declined sharply because of habitat destruction and the use of herbicides, but is still frequent on the East Anglian coast, the Brecklands of Norfolk and Suffolk, and the South Downs. The flowers are the colour of red wine, and the seeds are hard-coated sculptured nutlets.

Seed production: Hound's-tongue is easily grown from seed in small plots. The hard seeds are retained on the plant, which may be harvested whole, dried, and the seeds separated by passing through a combine harvester.

Seed of Hound's-tongue.

65. *Cynosurus cristatus*: Crested Dog's-tail (native)

Features: Crested Dog's-tail is a widespread, palatable, short-lived perennial grass of a wide range of grasslands on neutral to base-rich soils, and is present in grasslands throughout Britain and Ireland. It has a short, tufted, open growth form, with tall, thinly spaced inflorescences on a wiry stem, with each individual inflorescence cluster resembling a dog's tail. These features make Crested Dog's-tail the ideal, number one, companion native grass for restored and created meadows and other wild grasslands on most soils. Crested Dog's-tail was used in agricultural seed mixes until the 1940s, with seed coming from New Zealand. The origin of the New Zealand stock is thought to be native British, introduced by early settlers. The cultivar 'Southlands' remains available from New Zealand.

Seed production: Native wild British origin stock of Crested Dog's-tail is preferred, and much increased home production would be good. Large-scale row-crop production

Crested Dog's-tail (*Cynosurus cristatus*) at anthesis.

Seed of Crested Dog's-tail.

and harvesting is not problematic, and Crested Dog's-tail is not subject to the Seed Marketing Regulations 2011.

66. *Dactylis glomerata*: Cocksfoot (native)

Features: Cocksfoot is a common and widespread, strongly tufted, coarse perennial grass of a wide variety of habitats and soils across Britain and Ireland. It has been sown in agricultural leys for ages because it is palatable to livestock, but if left ungrazed or uncut it can quickly dominate a sward. It is not the best grass species for a meadow, but good for tussock grassland on most soils, providing structure in support of raptors and other fauna.

Seed production: Cocksfoot is a proscribed species subject to the Seed Marketing Regulations 2011, and is a forage grass popular with many farmers. Nevertheless, wild native stock has a place in the restoration of tussock and wayside grasslands. It is amenable to large-scale row-crop seed production, and seed yields are high.

67. *Dactylorhiza fuchsii*: Common Spotted-orchid (native)

Features: Common Spotted-orchid is the most widespread and most frequently encountered species of Orchid in Britain and Ireland. It is a tuberous polycarpic perennial of meadows, pasture, scrub, waysides, marshes, fens and

Cocksfoot (*Dactylis glomerata*) at anthesis.

Clonal group of Common Spotted-orchid (*Dactylorhiza fuchsii*).

Dust seed of Common Spotted-orchid.

wasteland, on neutral to calcareous substrates. It is characterised by lanceolate leaves that are usually spotted or blotched with black, and has a pinkish inflorescence spike that is rather pointed. It is a highly variable species, exhibiting variation in vigour, leaf markings, flower colour, and labellum shape and markings. Common Spotted-orchid is diploid and freely hybridises with most other *Dactylorhiza* species that it may be within pollinator distance of, some species of which may be tetraploid. Hybrids within ploidy level are very fertile, and even triploid hybrids between ploidy levels are usually sufficiently fertile to backcross to the parents. The most frequent hybrid in the wild is the triploid hybrid between diploid *D. fuchsii* and the tetraploid *D. praetermissa*, and such hybrids often generate a hybrid swarm by backcrossing and segregation. These hybrid swarms are usually characterised by hybrid vigour and considerable genetic variation, expressed in intermediate combinations of characters derived from both parents.

Seed production: Seeding spikes at the yellow pod stage may be collected from plants in cultivation, or from plants that are unavoidably about to be cut or destroyed. On drying, these spikes release buff-coloured dust seed that may be cleaned by passing through a flour sieve.

68. *Dactylorhiza praetermissa*: Southern Marsh-orchid (native)

Features: Southern Marsh-orchid is a tetraploid tuberous polycarpic perennial that may be found on slightly acid to alkaline soils in a wide range of habitats in the southern half of the British Isles. Its northern variant is the Northern Marsh-orchid (*Dactylorhiza purpurella*). Southern Marsh-orchid is characterised by leaves that are usually not spotted or blotched, an inflorescence spike that is rather blunt-tipped, and a flower colour that is a deeper pink/purple compared to that of Common Spotted Orchid. Use of seed of Southern Marsh-orchid should be restricted to the native range of the species in the southern half of the British Isles.

Southern Marsh-orchid (*Dactylorhiza praetermissa*).

Seed production: Seeding spikes at the yellow pod stage may be collected from plants in cultivation, or from plants that are unavoidably about to be cut or destroyed. On drying, these spikes release buff-coloured dust seed that may be cleaned by passing through a flour sieve.

69. *Daucus carota*: Wild Carrot (native)

Features: Wild Carrot is a mid-season biennial/winter annual Umbellifer of infertile, well-drained, neutral to calcareous soils. It is widespread in grasslands across the British Isles, while in Scotland it is mostly confined to the coast. Ecotypes exist, showing adaptations to inland and coastal conditions and other situations. A mystery remains regarding the function of the dark-red spot in the centre of the inflorescence, first noted by Darwin, and still debated

Wild Carrot (*Daucus carota*) flower with central red spot.

Seed of Wild Carrot.

(Polte and Reinhold, 2013; Raguso, 2023). One theory is that it imitates a blood spot to attract fly pollinators. Wild Carrot is a proscribed species subject to the Seed Marketing Regulations 2011. These regulations do not distinguish between the wild form of the species and the Garden Carrot, and therefore enable the supply of the Garden Carrot and prohibit the supply of the wild type.

Seed production: Wild Carrot seed has no dormancy restrictions and crops may be established from a late spring sowing for harvest in the following year. In cultivation, the wild type is very susceptible to Carrot Root Fly (*Chamaepsila rosae*). If this can be avoided seed yield can be very good, provided maturation and harvest time is storm free. The crop combines readily and can produce big yields.

70. *Deschampsia cespitosa*: Tufted Hair-grass (native)

Features: Tufted Hair-grass is a long-lived wintergreen perennial of slightly acidic to alkaline soils, and is able to grow in a wide range of conditions throughout Britain and Ireland. It is tolerant of shade and waterlogged soils and may be found in woodland, rough and wet grasslands, fen meadows, grass heath and a wide range of upland habitats. It is a rather large plant when mature, forming discrete upright tussocks of longish, wiry, dark green leaves. On wet ground it can form large tussocks, creating dry high spots

Tufted Hair-grass (*Deschampsia cespitosa*).

Seed of Tufted Hair-grass.

raised above wet ground. The high silica content of the leaves makes Tufted Hair-grass unpalatable to livestock, and this allows the plant to persist in overgrazed pastures.

Seed production: Tufted Hair-grass may be precision drilled in rows at a field scale. It grows well in cultivation and will produce multiple harvests, with the combine set with the drum half closed and the aspiration minimal.

71. *Digitalis purpurea*: Foxglove (native)

Features: Foxglove is a familiar biennial of hedge banks, open woodland, clearings, heathland and moorland margins, river-banks, sea-cliffs and upland rocky slopes, on neutral to acidic soils. It is frequent throughout Britain and Ireland, and is sometimes abundant following tree felling. The seed has good longevity, and germination is usually unhindered by dormancy.

Seed production: Foxglove grows well in cultivation and produces large amounts of fine seed. Stems may be gathered by hand at maturity pre-dehiscence, and drying on sheets releases the seed.

Foxglove (*Digitalis purpurea*) on Gower Peninsula.

Seed of Foxglove.

72. *Dipsacus fullonum*: Wild Teasel (native)

Features: Wild Teasel is a robust biennial with stems reaching 1.5–2.0m (5–6.5ft). It has spread since the 1960s from its former stronghold in the south-east of England to places across lowland England, Wales, southern and eastern Scotland, and many lowland regions in Ireland. Wild Teasel may be found growing in rough grassland, wood margins, scrub and hedgerows, riverbanks, wasteland and

Wild Teasel (*Dipsacus fullonum*).

Seed of Wild Teasel.

waysides. It is sometimes found growing with the less common Small Teasel (*Dipsacus pilosus*).

Seed production: In cultivation Wild Teasel is very easy to grow and produces abundant seed. Harvesting seed by hand is not easy as the plants and seed heads are covered in spines, but harvesting is very easily achieved with a combine harvester. Seed yield is good.

73. *Dryopteris felix-mas*: Male Fern (native)

Features: Male Fern is one of our commonest ferns and may be found throughout Britain and Ireland. It is a robust, shade-tolerant, rather grazing-intolerant, deciduous fern with spreading fronds, two to three pinnate, on soils that are moist and well drained, and slightly acidic to slightly basic. Male Fern may be found in woodlands, hedgerows, ditches, waysides, stream banks, rocky hillsides, cliff edges and scree slopes as well as many urban habitats. It self-propagates by air-borne spores and may be encouraged to do so by providing suitable habitat.

Spore collection: Male Fern is a species that may self-colonise a restored habitat, and should be welcomed if it does. Alternatively, sporulating fronds may be collected when the sori on the underside are mature.

74. *Echium vulgare*: Viper's Bugloss (native)

Features: Viper's Bugloss is a vivid blue-flowered biennial herb of grassy places and disturbed habitats on a wide

Male Fern (*Dryopteris felix-mas*).

Viper's Bugloss (*Echium vulgare*).

Seed of Viper's Bugloss.

range of well drained soils. It is a plant of slightly disturbed ground found growing on chalk and limestone downs, heaths, in quarries and chalk-pits, on waysides, cliffs, sand dunes, shingle banks, and on waste land.

Seed production: In cultivation, Viper's Bugloss grows well and a combine harvester will easily harvest the hard, angular seed, when ready. Beware the dust rising from plants at harvest time as it consists of fine irritant hairs.

75. *Epipactis helleborine*: Broad-leaved Helleborine (native)

Features: Broad-leaved Helleborine is the most frequent *Epipactis* species found in Britain and Ireland, and frequent throughout Scotland except in Northern Scotland where it is rare. A rhizomatous (rather than tuberous, as most Orchid species) perennial tolerant of a wide range of soil conditions and found in woodland, scrub and hedgerows. The flowers are pinkish green, and the labellum has a central depression containing a liquid that appears attractive to flies, which when visiting pick up and deposit pollinia, effecting cross-fertilisation. Flowering is in August and September and seed liberation in October.

Seed production: As with all Orchid species, seed must not be collected from plants in the wild, but may be obtained from plants in cultivation, or from plants about to be unavoidably destroyed.

76. *Epipactis palustris*: Marsh Helleborine (native)

Features: Marsh Helleborine is a declining rhizomatous perennial Orchid of diverse habitats including fens, marshes,

Broad-leaved Helleborine (*Epipactis helleborine*).

damp pastures, meadows, dune slacks, quarries, gravel pits, and occasionally chalk grassland, across much of Britain and Ireland. The few-flowered inflorescences are very attractive, and seed production and establishment from sown seed are both good. *The Biological Flora for Marsh Helleborine* has been published (Jacquemyn et al., 2014).

Seed production: Marsh Helleborine is a restoration species that may be used within its native range (only) using seed collected from plants in cultivation.

77. *Erica cinerea*: Bell Heather (native)

Features: Bell Heather is a key and common species of dry heaths and moors and other suitable habitat in Ireland, Scotland, Wales and England.

Seed production: Like Heather (*Calluna vulgaris*) and Cross-leaved Heath (*Erica tetralix*), the fine seed of Bell Heather is amenable to collecting by brush harvester from wild stands.

78. *Erica tetralix*: Cross-leaved Heath (native)

Features: Cross-leaved Heath is a common low shrub of wet heaths and moors throughout The British Isles.

Seed production: The fine seed of Cross-leaved Heath may be brush harvested from wild stands where conditions allow this, and where it is permitted to do so. The material obtained may be applied directly to restoration sites.

79. *Ervilia sylvatica*: Wood Vetch (native)

Features: Wood Vetch is an ascending perennial of hedges, woodland borders and clearings, in scrub and on ungrazed grassland on cliffs and screes. It is infrequent and locally distributed across Britain and Ireland.

80. *Eryngium maritimum*: SEA HOLLY (native)

Features: Sea Holly is a blue-flowered, very glaucous, spiny-leaved perennial restricted to coastal habitats. It is mostly found growing on mobile sand dunes and occasionally on shingle. *The Biological Flora for Sea Holly* has been published (Isermann and Rooney, 2014).

Seed production: Sea Holly may be grown in cultivation in small plots using horticultural methods. At seeding, and wearing leather gloves, the seed heads can be detached with scissors; on drying, the large spine-fringed seeds can be separated by sieving.

81. *Eupatorium cannabinum*: Hemp Agrimony (native)

Features: Hemp Agrimony is a tall, mid- to late summer-flowering perennial found growing across a wide range of moist and wet habitats on calcareous soils, where it attracts pollinating butterflies. The name Hemp Agrimony is derived from the Hemp-like leaves of the species. It is common across most of the southern half of Britain and Ireland, and may be found growing at the edge of ponds, lakes, rivers and canals. It is also found in marshes, wet meadows, tall-herb

Sea Holly (*Eryngium maritimum*).

Hemp Agrimony (*Eupatorium cannabinum*).

Seed of Hemp Agrimony.

fen, wet woodland, mires and wet heaths, on hedge banks and flushed areas of sea cliffs and dune slacks. Hemp Agrimony produces seed with a pappus, and the fluffy seed is retained on the inflorescence into late summer.

Seed production: Hemp Agrimony seed may be collected from where it may be found growing, provided it is permitted to do so. Collection of the seed heads is easy, but the pappus is directly and strongly attached to the seed and is difficult to remove.

82. *Euphorbia exigua*: Dwarf Spurge (archaeophyte)

Features: Dwarf Spurge is an uncommon spring annual of arable cultivation, formerly much more common and widespread across Britain and Ireland. It is a spring annual of sandy and calcareous soils, forming attractive fine-leaved cushions with greenish-yellow flowers. Outside arable cultivation it may occasionally be found on disturbed soil in calcareous grassland, in gardens, on wasteland and in churchyards.

Dwarf Spurge (*Euphorbia exigua*).

83. *Euphorbia helioscopia*: Sun Spurge (archaeophyte)

Features: Sun Spurge is an attractive and frequent spring annual of disturbed ground and arable cultivation on well-drained, neutral to calcareous soils. Like most *Euphorbia* species, broken stems can bleed a white fluid, to which some people may be sensitive, causing a skin reaction.

Flower cluster of Sun Spurge (*Euphorbia helioscopia*).

84. *Festuca ovina*: Sheep's Fescue (native)

Features: Sheep's Fescue is a densely tufted, fine-leaved, non-rhizomatous perennial grass on soils ranging from calcareous to acid. It is an important constituent of unimproved pastures and meadows, and other grasslands across much of Britain and Ireland. Sheep's Fescue is often the main constituent of calcareous and acid grasslands, dry heathlands, rocky ledges and sea-cliffs, and of other grasslands on well-drained and infertile habitats. Sheep's Fescue is a proscribed species under the Seed Marketing Regulations 2011, and cultivars are used in amenity grass

Seed of Sheep's Fescue.

seeding. Sheep's Fescue is a variable species and is present in the wild as numerous locally adapted ecotypes.

Seed production: Field-scale seed production of wild stock of Sheep's Fescue, although not permitted by the regulations, is straightforward using standard agricultural methods, and seed yields are very good.

85. *Festuca rubra*: Red Fescue (native)

Features: Red Fescue is a very variable perennial grass found throughout Britain and Ireland. It is a fine-leaved grass and may be tufted in habit, or rhizomatous. Wide-ranging ecotypes of Red Fescue can be found in all types of grassy habitat, from lowland meadow and salt marshes to mountain grasslands and rocky ledges. Red Fescue is a proscribed species under the Seed Marketing Regulations 2011, and cultivars are used to produce fine-leaved lawns and for general amenity landscaping.

Seed production: Both Sheep's Fescue and Red Fescue are very important restoration species, and field-scale seed production of wild stock, although not permitted by the regulations, is straightforward using standard agricultural methods, and seed yields are very good.

86. *Filipendula ulmaria*: Meadowsweet (native)

Features: Meadowsweet is a tall, sweet-smelling, summer-flowering perennial of moist or seasonally wet ground. It grows on neutral and calcareous soils in meadows, along riversides, waysides, in swamps, tall-herb fens and in montane communities. The leaves are once pinnate and toothed, and the flowers are cream-coloured. It is a characteristic species of floodplain meadow MG4 communities, and is common in the wild throughout Britain and Ireland. Meadowsweet, formerly known as *Spiraea ulmaria* (because it has sickle-shaped seeds arranged spirally), is a natural source of salicylic acid. This is the drug we know as 'aspirin', the name being derived from 'of Spiraea', so we have Meadowsweet to thank for aspirin. It is also an important restoration species.

Red Fescue (*Festuca rubra*).

Seed of Red Fescue.

Meadowsweet (*Filipendula ulmaria*).

Seed of Meadowsweet.

Seed of Dropwort.

Seed production: The autumn-drilled seed requires vernalisation to break dormancy; Meadowsweet establishes and cultivates easily on moist ground. Seed crops can be combine harvested at light-brown seed stage, yielding well. Seed may also be hand collected from wild stands (where it is permitted to collect seed), doing this as the seeds turn brown, but not earlier. The seed is retained well on the plant, and benefits from a later harvest.

87. *Filipendula vulgaris*: Dropwort (native)

Features: Dropwort is an early summer-flowering perennial of chalk and limestone grasslands and some neutral lowland meadows. It is widespread, but declining, in Britain and Ireland, with flowers that are pink in bud, opening to white. The species has much divided basal leaves that are twice pinnate.

Seed production: Dropwort is an important restoration species and crops may be grown on calcareous land. The seeds may be autumn drilled in rows, requiring vernalisation to break dormancy, and crop establishment is slow. When established, crops yield well, although deer are inclined to browse flowers. Harvest may be by hand or combine, at the light-brown seed stage.

Dropwort (*Filipendula vulgaris*).

88. *Fragaria vesca*: Wild Strawberry (native)

Features: Wild Strawberry is a stoloniferous perennial of neutral to calcareous soils and may be found growing in open woodland, scrub, hedgerows and waysides across much of the British Isles. When not in fruit, it may be confused with the slightly more pubescent Barren Strawberry (*Potentilla sterilis*). Wild Strawberry is diploid (2n=2x=14) and is not involved in the parentage of the octoploid (2n=8x=56) Garden Strawberry (*Fragaria* × *ananassa*).

Seed production: The fruit of Wild Strawberry may be collected by hand and the seeds extracted by pulping and

Wild Strawberry (*Fragaria vesca*).

sieving, or by drying and rubbing the seeds from the fruit surface. Seed yield will, of course, be very low. Propagation will often be easier by raising plants from rooted stolons.

89. *Fritillaria meleagris*: Snake's-head Fritillary (naturalised neophyte)

Features: Snake's-head Fritillary is a spring-flowering bulbous perennial naturalised in floodplain meadows. It was recorded in cultivation in the British Isles by 1596, and was first recorded in the wild in 1736 (Walker, 2021). Snake's-head Fritillary is the iconic species of Thames Valley floodplain meadows, and around 80 per cent of the national population resides on North Meadow National Nature Reserve (NNR), a short walk from the centre of Cricklade. *The Biological Flora of Snake's-head Fritillary* is published (Tatarenko et al., 2022).

Snake's-head Fritillary (*Fritillaria meleagris*) at Cricklade, mostly dark pink flowered.

Snake's-head Fritillary (*Fritillaria meleagris*) at Ashton Keynes, mostly white flowered.

Seed of Snake's-head Fritillary.

Seed production: The seed of Snake's-head Fritillary requires vernalisation to break dormancy; it may be sown in the field in the autumn, or plants may be grown from seed under glass and planted out. Harvest is in June, and pods can be collected by hand just before they split, then dried to release the seed.

90. *Fumaria officinalis*: Common Fumitory (archaeophyte)

Features: Common Fumitory is an uncommon spring annual of cultivated and disturbed ground on calcareous and light soils. It is an attractive, low-growing species with divided leaves and pinkish flowers. The round seeds are foraged by birds, and if they fall to the ground they can lie dormant for many years.

Common Fumitory (*Fumaria officinalis*).

Seed production: In cultivation, Fumitory will produce good quantities of seed in the following summer from an autumn or early spring drilling; the seed may be harvested directly with a combine harvester.

91. *Galeopsis tetrahit*: Common Hemp-nettle (native)

Features: Common Hemp-nettle is an upright annual of cultivated and disturbed soils across most of the British Isles. It may be seen growing in woodland clearings, ditches, fens, waysides, wet heaths and along riverbanks.

Seed production: Common Hemp-nettle may be precision drilled in the autumn and the seed crop harvested by combine in the following summer.

Common Hemp-nettle (*Galeopsis tetrahit*).

92. *Galium album*: Hedge Bedstraw (native)

Features: Hedge Bedstraw (former Latin name *Galium mollugo*) is a wintergreen rhizomatous perennial that is native to Britain but not Ireland, where it occurs as a

Hedge Bedstraw (*Galium album*).

Seed of Hedge Bedstraw.

naturalised neophyte. Hedge Bedstraw grows in well-drained calcareous to neutral soils and is mostly a lowland plant, being common and widespread. It is frequent on waysides, hedge banks and among scrub, with long, clambering stems bearing clusters of four-petalled small white flowers. Hedge Bedstraw is equally at home in calcareous grassland exposed to full sun.

Seed production: Hedge Bedstraw is a very easy species to grow in cultivation and will crop for several years producing large amounts of seed. Harvesting is by combine harvester.

93. *Galium odoratum*: Sweet Woodruff (native)

Features: Sweet Woodruff is a white-flowered, fragrant rhizomatous perennial of deciduous woods and shaded hedgerows, found infrequently on calcareous and often damp soils throughout the British Isles. It is highly

Sweet Woodruff (*Galium odoratum*).

rhizomatous and will often form a large, cushion-like patch in woodland, showing off characteristically dark-green, semi-glossy leaves in whorls.

Seed production: Sweet Woodruff is not an easy species to cultivate in the open and seed production is often disappointing. It is most easily propagated by root division.

94. *Galium verum*: Lady's Bedstraw (native)

Features: Lady's Bedstraw is a well-loved native of freely draining, infertile, neutral to calcareous soils in grassy places, and is frequent and widespread across the British Isles. It is a yellow-flowered, fragrant, stoloniferous, fine-leaved, deciduous perennial of hay meadows and pastures, coastal cliff-tops, dune grasslands, waysides and machair. Flowering is in July, rather late for a normal hay cut, but Lady's Bedstraw can readily survive vegetatively from year to year without setting seed. It is a variable species and exists as different ecotypes in different habitats, and sometimes hybrids between different

Lady's Bedstraw (*Galium verum*).

Seed of Lady's Bedstraw.

Galium species may be found. In North America, Lady's Bedstraw is considered to be an emerging invasive alien of native Tall Grass Prairie (Heckman et al., 2016).

Seed production: Lady's Bedstraw may be precision drilled in rows in the autumn or spring, at a field scale. In cultivation it is a species that grows well, but it is very susceptible to Powdery Mildew in dry weather, especially at flowering and after. The seed crop may be combine harvested, and in a good year yields are very respectable, although a little less than that of Hedge Bedstraw.

95. *Genista tinctoria*: Dyer's Greenweed (native)

Features: Dyer's Greenweed is a small, slender leguminous shrub of slightly acid to calcareous grasslands. It is a plant of old meadows, pastures and other grassy places, and has declined considerably because of agricultural improvement, under grazing, neglect, and also, more recently, overgrazing with sheep. It is more frequent in the south, and is absent from central and northern Scotland and Ireland. Dyer's Greenweed is a host plant for many important and rare invertebrates (Stroh, 2015b).

Dyer's Greenweed (*Genista tinctoria*).

Seed production: Dyer's Greenweed is suitable for small-scale cultivation. It is slow to establish from seed, but is a long-lived perennial and will continue to yield successive seed harvests. Harvest may be by hand collection of mature pods and drying to release the seed.

96. *Geranium dissectum*: Cut-leaved Crane's-bill (archaeophyte)

Features: Cut-leaved Crane's-bill is a common and widespread annual of grassland, cultivated ground and disturbed ground on waysides, hedge banks, in gardens and many other places. The lobed leaves are deeply dissected, and the small pink flowers produce seed that is explosively liberated.

Seed production: Cut-leaved Crane's-bill is a vigorous annual and may be autumn or spring drilled in rows. The

Cut-leaved Crane's-bill (*Geranium dissectum*).

seeding plants are best cut at maturity and placed on a sheet to dry. The released seed may be swept up, and the remaining material passed through a combine harvester. Seed yield can be reasonable.

97. *Geranium lucidum*: Shining Crane's-bill (native)

Features: Shining Crane's-bill is an annual of hedge banks, walls and rocky places on calcareous soils. The lobed leaves are not deeply divided and are glossy, at first green, later turning reddish.

Seed production: The small pink flowers of Shining Crane's-bill produce seed that is explosively liberated and difficult to collect. It is a species that is best grown in small plots with the seed harvested by hand. Only small quantities will be obtained in this way.

Shining Crane's-bill (*Geranium lucidum*).

98. *Geranium molle*: Dove's-foot Crane's-bill (native)

Features: Dove's-foot Crane's-bill is a prostrate, cushion-forming and softly pubescent annual occurring in a wide variety of open habitats including dry grasslands and rocky outcrops, garden lawns and cultivated land. The lobed leaved are moderately dissected and similar to, but smaller than, the leaves of Hedgerow Crane's-bill. The small flowers are pink, or sometimes white.

Seed production: Cultivation and seed collection is very difficult for Dove's-foot Crane's-bill. It is best grown in small plots with the seed harvested by hand.

Pale-flowered variant of Dove's-foot Crane's-bill (*Geranium molle*).

99. *Geranium pratense*: Meadow Crane's-bill (native)

Features: Meadow Crane's-bill is one of our best-loved native species. It is a deciduous perennial and is native to central, southern and northern England, eastern Wales, and all but the extreme north of Scotland. Elsewhere it is introduced. Although widespread, it is not a frequent meadow species, as the name suggests, being a little too late in seed set to benefit from normal mowing dates. It prospers better on less intensively managed habitats such as waysides, hedge banks, scrub and wasteland. It is competitive in these habitats and can tolerate a wide range of soil types and levels of fertility.

Seed production: Meadow Crane's-bill is robust in cultivation. The seed requires scarification to break dormancy, and is best autumn drilled. Drill at 1cm (0.4in) depth, as pigeons will otherwise graze the seed. At flowering, the

Seed crop of Meadow Crane's-bill (*Geranium pratense*) at full flower.

Meadow Crane's-bill flower pedicels reflexed in bud and erect in full flower.

Meadow Crane's-bill flower pedicels reflexed after pollination.

Meadow Crane's-bill flower pedicels erect at seed maturation and dehiscence.

Meadow Crane's-bill seed crop ready for harvest.

Seed of Meadow Crane's-bill.

inflorescence pedicels dance, down/up/down/up. They droop in bud, are upright in flower, droop again during seed development, and are upright again as seeds mature. Mature seeds are dark brown and are held in a catapult-like mechanism that explosively throws the seed 1 to 3m (3 to 10ft). This is achieved by differential drying (and shortening) of the outer and inner surfaces of the beak segments until tension is sufficient to break the seed capsule free, resulting in explosive coiling of the beak segment and ballistic discharge of the seed. This has implications for seed harvest, because mature seed is not retained on the plant during hot/dry periods of low relative humidity, and direct combine harvesting can be poorly rewarded. Yield may be improved by allowing the crop to accumulate mature seed during periods of damp/cool weather and combine harvesting at the point just before the weather changes to hot/dry. The yield of seed can reduce by more than half as the weather changes. Alternatively, in prolonged dry/hot weather, the crop can be combine harvested at peak yellow seed stage and the seed cured by very slow drying. Alternatively, smaller crop areas may be planted through woven polypropylene sheeting and the seed collected by sweeping. *The Biological Flora for Meadow Crane's-bill* has been published (Jefferson et al., 2023).

100. *Geranium pyrenaicum*: Hedgerow Crane's-bill (naturalised neophyte)

Features: Hedgerow Crane's-bill is not native to our islands, but has been recorded in the wild in Britain since 1762. It is now well naturalised and is locally frequent in hedgerows, waysides and in rough, grassy places. Hedgerow Crane's-bill is a wintergreen biennial/short-lived perennial with purplish pink flowers and large, rounded leaves. It is not invasive and is a welcome addition to the wild hedgerow flora.

Hedgerow Crane's-bill (*Geranium pyrenaicum*).

Seed of Hedgerow Crane's-bill.

Seed production: Hedgerow Crane's-bill may be precision drilled in rows at a field scale. It grows very well in cultivation and the seed may be harvested directly by combine harvester in the year after drilling. Seed yield is very good for a *Geranium* species.

101. *Geranium robertianum*: Herb Robert (native)

Features: Herb Robert is a common annual of hedgerows, waysides and rocky places throughout Britain and Ireland. The small flowers are pink with darker pink venation, and the leaves are much divided. Rubbing the leaves releases a characteristic musty aroma.

Herb Robert (*Geranium robertianum*).

Seed production: Herb Robert can sometimes be plentiful on abandoned hard or rubble surfaces, providing an opportunity to collect seeding plant material, where this is permitted. When placed on a sheet to dry, the seeds dehisce in the sun, being thrown a short distance, and can be collected from the sheet. The yield is always very small, and in cultivation it is no better.

102. *Geum rivale*: Water Avens (native)

Features: Water Avens is a rhizomatous perennial of moist soils. It is found throughout Britain and Ireland on canal sides and riverbanks, wet flushes, damp hay meadows, and at higher altitudes in montane willow scrub and amongst tall herbs on ledges. The pretty, nodding, pinkish flowers produce small bunches of fluffy seeds.

Seed production: Water Avens is best suited to small-scale row cultivation using horticultural methods; seed harvest is normally by hand collection.

Seed of Water Avens.

103. *Geum urbanum*: Wood Avens (native)

Features: Wood Avens is a common perennial of woodland, hedges and waysides. The small yellow flowers produce clusters of seeds with a hooked appendage. The hybrid between Water Avens and Wood Avens is intermediate in form.

Seed production: Wood Avens may be precision drilled in rows at a field scale. It is a species that grows well in cultivation, producing abundant seed that is retained on the plant, and which is easily harvested by combine harvester.

Wood Avens (*Geum urbanum*).

Seed of Wood Avens.

104. *Glaucium flavum*: Yellow-horned Poppy (native)

Features: Yellow-horned Poppy is a short-lived, infrequent, perennial halophyte of shingle and stony beaches confined to coastal locations.

Seed production: Yellow-horned Poppy may be established from seed, and is best cultivated on a small scale at first. The long seedpods are hand harvested just before splitting lengthways, and are dried to release the seed.

105. *Glebionis segetum*: Corn Marigold (archaeophyte)

Features: Corn Marigold (former Latin name *Chrysanthemum segetum*) is a frost-tender spring annual of cultivated fields on neutral to slightly acid, loamy and sandy soils. It is intolerant of calcareous soils. Once banned by Henry II, it was a serious weed of wheat and potatoes in Victorian times. Now it is much less common than it was then, due to a move to autumn cropping, improved seed cleaning, liming, and the use of herbicides. Corn Marigold is still locally abundant in some isolated fields on light soils along the coast. It is an attractive species with large, yellow, daisy-like flowers and glaucous leaves. The seed has good longevity in the soil seed bank, and is polymorphic, with ray florets producing winged seed and disc florets producing ribbed wingless seeds.

Corn Marigold (*Glebionis segetum*).

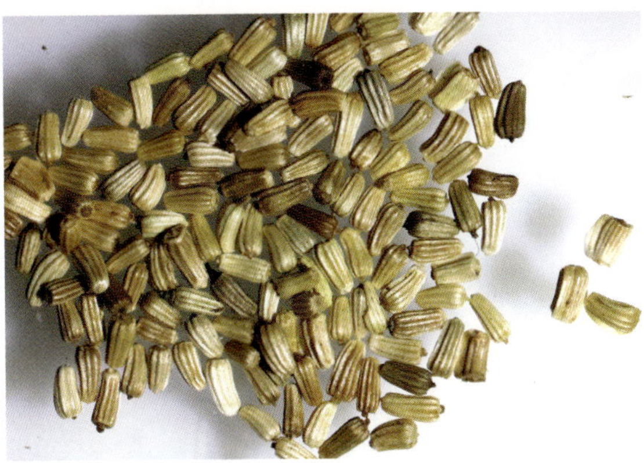

Seed of Corn Marigold.

Seed production: Corn Marigold is an important restoration species for creating annual communities. Precision drill in rows at a field scale in the spring, and combine harvest in the summer at seed maturity.

106. *Helianthemum nummularium*: Rockrose (native)

Features: Rockrose is a mat-forming, low-growing subshrub of nutrient-poor thin soils on calcareous to mildly acid substrates throughout Britain and Ireland. The flowers are attractively yellow, and the small glossy leaves are wintergreen.

Seed production: Although Rockrose is an important constituent of our native flora it is not an easy species to grow in cultivation, or from which to harvest seed. It is best suited to small plot cultivation using horticultural methods and hand harvest.

Rockrose (*Helianthemum nummularium*) on Gower Peninsula.

107. *Heracleum sphondylium*: Hogweed (native)

Features: Hogweed is an unjustly maligned, robust biennial of meadows, hedgerows, waysides and rough ground on neutral to calcareous soils throughout Britain and Ireland. The whole plant is highly palatable to all livestock, it provides abundant pollinator services, and standing dead in winter the hollow stems fill with hibernating invertebrates. In a meadow, Hogweed provides welcome structure, and numbers will naturally wax and wane, rarely dominating.

Seed production: Cultivation for seed is not necessary as Hogweed is such a common plant of waysides and farmland. Hand collection is easy, or plants may be combine harvested *in situ* on the farm. Hogweed deserves a more prominent role in restoration than it has had in the past.

Hogweed (*Heracleum sphondylium*) provides nourishment for aphids and ladybird larvae.

Seed of Hogweed.

108. *Hippocrepis comosa*: Horseshoe Vetch (native)

Features: Horseshoe Vetch is an uncommon and declining, mat-forming, sweet-scented perennial legume of dry, sunny and open pastures on chalk and limestone. It is also found on chalk and limestone cliffs, and shingle. In flower it produces abundant clusters of yellow flowers that resemble those of Bird's-foot Trefoil (*Lotus corniculatus*). In seed it

Horseshoe Vetch (*Hippocrepis comosa*).

Single-seeded pod segments of Horseshoe Vetch.

produces segmented pods that break into single-seeded segments that are curved like a horseshoe. Horseshoe Vetch produces carpets of yellow flowers and scores very high on ecosystem services, especially for pollinator services and as the host for life cycle stages of numerous invertebrates. Horseshoe Vetch is present in Britain as two cytotypes: a diploid race (2n=2x=14) that is confined to ungrazed hard limestones of the south and west of England and Derbyshire, and a much more widespread tetraploid race (2n=4x=28) that occurs throughout the range of the species (Fearn, 1972).

Seed production: Horseshoe Vetch grows well in cultivation on calcareous soils and produces abundant seed, with pods retained on the plant well after maturity. If grown with a moderate density of Creeping Thistle (*Cirsium arvense*), the seeding inflorescences of Horseshoe Vetch are raised sufficiently high to allow harvesting by combine harvester. Otherwise, pods can be hand harvested. The seed can be separated from the pod casing, but this is not always necessary.

109. *Hordeum secalinum*: Meadow Barley (native)

Features: Meadow Barley is an occasional perennial grass of meadows, pastures and roadsides on clay soils. It is similar in appearance to Wall Barley (*Hordeum murinum*), a familiar annual of pavement and pathway sides.

Seed production: Meadow Barley may be precision drilled in rows at a field scale to produce seed crops that can be harvested by combine harvester. The seed is awned, and the awns can be removed to assist cleaning and sowing, if necessary.

110. *Hyacinthoides non-scripta*: Bluebell (native)

Features: Bluebell is a bulbous perennial of deciduous woodland, and although declining, it remains frequent to abundant throughout the British Isles. It also grows in hedgerows, waysides, shady banks, coastal grasslands and meadows. Our native species hybridises freely with the garden Spanish Bluebell (*Hyacinthoides hispanica*) to produce the hybrid *Hyacinthoides* x *massartiana*. Native Bluebell in flower is a spring feature of the short grassland on Brean Down in Somerset, the Bluebell meadows of the Mendips, and woodlands countrywide. Good seed set depends on good flight conditions for early emerging pollinators, and the seeds, when produced, are black, round, and slightly glossy, and are held loosely in papery capsules. The seeds require vernalisation to break dormancy and will germinate in the spring. The seedlings soon produce a minute proto-bulb at

Bluebell (*Hyacinthoides non-scripta*).

Seed of Bluebell.

the point where the shoot and the root meet, and bunches of contractile roots extend into the soil and then contract, concertina-like, pulling the proto-bulb well below the surface to its perennial resting position.

Seed production: Native Bluebell is a protected species listed on Schedule 8 of The Countryside and Wildlife Act 1981, and wild collected seed may not be marketed, except when collected under licence from Natural England. If producing Bluebell seed in cultivation, the crop may be established from wild native origin seed, and the seed produced by that crop may be freely harvested and used. Harvesting will normally be by hand, gathering inflorescences as the seed turns from white to black, then drying these to release the seed. Drying can be rather slow because of the gelatinous nature of the seedpods. The digging and marketing of wild origin Native Bluebell bulbs is not permitted under any circumstances.

111. *Hypochaeris radicata*: Cat's-Ear (native)

Features: Cat's-ear is a widespread rosette-forming Dandelion-like perennial of grasslands. It is distinguished by having a branched, mostly hairless inflorescence. The seeds are dark reddish-brown in colour, narrow and rather pointed at both ends, and have a whitish pappus mounted on a narrow beak as long as the seed.

Seed production: Cat's-ear is difficult to grow and harvest in cultivation. It can be precision drilled in rows in the field and initial establishment is good, but it struggles to compete with weeds. Seed production is rather indeterminate, and mature seed is easily dislodged by wind. Seed harvest is usually by hand, and the pappus is easily removed by rubbing. Samples from brush-harvested meadows will often contain seed of Cat's-ear, but sorting out a clean sample of Cat's-ear seed is near impossible.

Cat's-ear (*Hypochaeris radicata*).

112. *Iris foetidissima*: Gladdon (native)

Features: Gladdon is a superbly handsome, rhizomatous, long-lived perennial of woods, hedgerows, scrub and sea

Gladdon (*Iris foetidissima*).

Seed of Gladdon.

cliffs on calcareous soils. It has mauve flowers with dark veins and bright orange fleshy seeds that are distributed by birds. Originally found only in southern Britain, its range appears to be extending northwards.

Seed production: Gladdon may be established in cultivation by the sowing of seed or by planting rhizome divisions taken from existing plants in cultivation. It is a species that grows well, and will produce good quantities of seed for many years. The seedpods are best harvested by hand after they have split to expose the red fleshy seeds (but not before). Drying will take a few months to complete in ambient conditions.

113. *Iris pseudacorus*: Yellow Flag-iris (native)

Features: Yellow Flag-iris is a robust, rhizomatous, long-lived perennial frequently found growing in marshes, wet woods, wet meadows, along watercourses, around lakes and ponds, in wet dune-slacks, the upper reaches of salt marshes and other wet coastal habitats. The yellow flowers are large, and they mature to produce green seedpods that split lengthways to reveal large brownish/orange seeds. The testa does not shrink as the seeds dry and mature, and the resultant light seeds float on water and are dispersed in that way.

Yellow Flag-iris (*Iris pseudacorus*).

Seed of Yellow Flag-iris.

Seed production: The seedpods of Yellow Flag-iris are easily collected by hand from existing populations after the pods have split (but not before), and where it is permitted to do so. Drying of the seed is gradual, and will take many weeks in ambient conditions to complete. Harvested seed may be used to establish crops in cultivation, and rhizome division can further propagate these plants.

114. *Jasione montana*: Sheep's-bit (native)

Features: Sheep's-bit is a declining biennial of lowland heath, sea cliffs, maritime grasslands, stabilised dunes, and inland on walls and banks. It is a difficult species to grow in cultivation, and the attractive blue flowers produce only very small amounts of seed.

Seed production: Sheep's-bit seed may be germinated under glass and plants pricked out for growing on and eventual planting in the field in short rows. It may also be precision surface drilled in short rows directly in the field. The seed is so small that it is best mixed with a carrier for drilling. Cooked commercial *Agrostis capillaris* seed is an

Sheep's-bit (*Jasione montana*).

ideal carrier for fine seed. Mature plants of Sheep's-bit produce clusters of blue flowers that turn brown in seed. Seeding clusters can be collected by hand and dried to release the fine seed.

115. *Kickxia elantine*: Sharp-leaved Fluellen (archaeophyte)

Features: Sharp-leaved Fluellen is an uncommon annual of cultivated and disturbed ground found on calcareous and sandy soils. The small, attractive flowers are borne in the leaf axils of horizontally spreading, usually unbranched, stems.

Seed production: Sharp-leaved Fluellen is a difficult species to grow and obtain seed from, but it may spring from the soil seed bank following cultivation or disturbance. If this happens, seeding stems can be picked as pods mature, and dried to release the seed.

116. *Kickxia spuria*: Round-leaved Fluellen (archaeophyte)

Features: Round-leaved Fluellen is a little more widely occurring than Sharp-leaved Fluellen, but often the two species will be found together.

Seed production: Seed may be obtained in the same way as described for Sharp-leaved Fluen, above.

117. *Knautia arvensis*: Field Scabious (native)

Features: Field Scabious is a deep-rooted, long-lived perennial of neutral to calcareous soils. It is occasional and widespread on rough pasture, waysides, banks, hedgerows, and sometimes meadows. Wild populations can exhibit considerable genetic variation, and are responsive to changing environments (Rottstock et al., 2016). The blue flowers are highly attractive to butterflies and bees, and the pollinated flowers set seed arranged bunched on an

Field Scabious (*Knautia arvensis*).

Seed of Field Scabious.

exposed capitulum. *The Biological Flora for Field Scabious* has been published (Varga et al., 2022).

Seed production: Flowering of Field Scabious is rather indeterminate, and the best seed yields from small plots are obtained from repeated hand picking over several weeks. This is achieved by turning and plucking the individual seed heads at maturity, judged as being the point at which seeds are swollen and easily rubbed off. Field Scabious may also be grown at a field scale, precision drilled in rows. It grows robustly and is an upright crop, presenting well for combine harvesting at peak seed maturity.

118. *Koeleria macrantha*: Crested Hair-grass (native)

Features: Crested Hair-grass is an uncommon but widespread perennial grass of infertile calcareous grassland. It is less common than formerly because of habitat loss and agricultural intensification.

Seed production: Crested Hair-grass may be grown as a row crop on a field scale, and the seed is suitable for combing harvesting.

Yellow Archangel (*Lamiastrum galeobdolon*).

119. *Lamiastrum galeobdolon*: Yellow Archangel (native)

Features: Yellow Archangel is an attractive wintergreen native species of woods, hedges and waysides. It is stoloniferous, and propagates easily by digging rooted stolons. Seed production in cultivation is not easy. Also, beware Variegated Yellow Archangel (*Lamiastrum galeobdolon ssp argentatum*): this is a seriously invasive alien garden escapee/throw-out, harming wayside and woodland native biodiversity; it is a 'prohibited species' listed on Schedule 9 of The Countryside and Wildlife Act 1981. On no account should Variegated Yellow Archangel be propagated or introduced to the wild.

Seed production: Yellow Archangel propagates easily from rooted stolons, but less readily by seed. Despite the very attractive flowers, seed production is rather poor.

White Dead-nettle (*Lamium album*).

120. *Lamium album*: White Dead-nettle (archaeophyte)

Features: White Dead-nettle is a most attractive plant, and it does not sting. It is a stoloniferous, sometimes rhizomatous, wintergreen perennial of wayside, waterside, woodland edge and on wasteland in village and town. White Dead-nettle is common and widespread in Britain, and local in Ireland.

Seed production: White Dead-nettle propagates easily from rooted stolons, but less easily by seed.

121. *Lamium purpureum*: Red Dead-nettle (archaeophyte)

Features: Red Dead-nettle is a very common and widespread annual of fertile and disturbed soils. It is a variable species, and so common that it is likely to spring from the soil seed bank whenever ground is cultivated or disturbed. Along with White Dead-nettle, both species are good at providing pollinator services.

Seed production: Red Dead-nettle is common on cultivated ground and seed may be easily collected by hand wherever the species is found.

Red Dead-nettle (*Lamium purpureum*).

122. *Lapsana communis*: Nipplewort (native)

Features: Nipplewort is a pale-yellow flowered annual/biennial composite of woodland, hedgerow, scrub and wayside. It is an inconspicuous but very common native plant, and its name derives from the form of the flowers in seed.

Seed production: Nipplewort is common, and seed may be easily collected by hand from waysides.

Nipplewort (*Lapsana communis*).

123. *Lathyrus pratensis*: Meadow Vetchling (native)

Features: Meadow Vetchling is a rhizomatous, yellow-flowered, climbing perennial legume with an angled stem. It is occasional in hedges and field borders, verges, hay meadows and unimproved pastures on moderately fertile soil.

Meadow Vetchling (*Lathyrus pratensis*).

Seed of Meadow Vetchling.

Seed production: In cultivation, a well established seed crop of Meadow Vetchling may be combine harvested to produce reasonable quantities of seed. The seed requires scarification to break dormancy.

124. *Lathyrus sylvestris*: Narrow-leaved Everlasting-pea (native)

Features: Narrow-leaved Everlasting-pea is a pink/purple-flowered scrambling perennial of wood edges, scrub, rough banks and sheltered sea cliffs. It is good at providing pollinator services and supports a wide range of invertebrates including the Shrill Carder Bee (*Bombus sylvarum*), our rarest bumblebee. Narrow-leaved Everlasting-pea is widespread, but not common, throughout Britain in the wild, and is introduced in many wild places.

Seed production: Narrow-leaved Everlasting-pea has been rarely used in restoration, but it is a native species that would cultivate easily, and produces seed that is easily harvested by a combine harvester.

125. *Legousia hybrida*: Venus's Looking-glass (archaeophyte)

Features: Venus's Looking-glass is a rare annual of cultivation and disturbed ground on calcareous soils. It is an inconspicuous plant with small lilac/purple flowers arranged in terminal cymes. The fine seeds are brown with a very glossy finish to the testa, resembling a mirror, and the seeds are long lived in the soil seed bank.

Seed production: Venus's Looking-glass seed is best suited to small-scale horticultural production methods.

Narrow-leaved Everlasting-pea (*Lathyrus sylvestris*).

Venus's Looking-grass (*Legousia hybrida*).

126. *Leontodon hispidus*: Rough Hawkbit (native)

Features: Rough Hawkbit is a widespread, rosette-forming, long-lived perennial of neutral to calcareous soils and is frequent on waysides and downland and in meadows, pasture, amenity grass and coastal grasslands. It is frequent across Britain and Ireland, but rare in Northern Ireland and northern Scotland. The yellow Dandelion-like flowers are borne singly on rather hairy stems, and these flowers mature to produce a head of spreading seeds, each having a pappus of buff-coloured feathery hairs. Unlike Dandelion and Cat's Ear, the pappus of Rough Hawkbit is attached directly to the seed, rather than to a beak extension of the seed.

Rough Hawkbit (*Leontodon hispidus*).

Seed of Rough Hawkbit.

Seed production: Rough Hawkbit is an important restoration species and field-scale cultivation is easy. Seed harvesting is not easy, and requires periods of still weather and specialist machinery. Once harvested, the pappus must be removed to allow for the seed to be cleaned.

127. *Leucanthemum vulgare*: Moon Daisy (native)

Features: Moon Daisy (Ox-eye Daisy) is a common and widespread short-lived perennial of meadows, pastures and waysides on neutral to calcareous soils. It is a variable

Moon Daisy (*Leucanthemum vulgare*).

Seed of Moon Daisy.

species and similar to the much larger-flowered garden Shasta Daisy (*Leucanthemum* x *superbum*), which is occasional in the wild as a naturalised neophyte. Moon Daisy is an important restoration species because the seeds have no dormancy, the plants establish readily on a wide range of substrates, and they flower, often in profusion, within two to three years of sowing. Inevitably populations will later decline in a created or restored meadow. Second and later generation recruitment may be encouraged by moderate poaching or scarification in winter, creating sites for establishment from self-sown seed.

Seed production: In cultivation, Moon Daisy grows well following precision drilling and produces abundant seed. Optimum time to harvest may be judged by peak turning of the white ray ligules to edged light brown and drooping. The seed is fine, ribbed and silver in colour. It may be combine harvested with the aspiration turned down, harvesting both the seed and the floret remnants. The seed can be cleaned from the florets, with care, to produce a good seed sample.

128. *Linaria vulgaris*: Toadflax (native)

Features: Toadflax is an occasional rhizomatous perennial of waysides and wasteland throughout England and Wales. It is narrow leaved, and the yellow flowers have two orange-coloured lips and a yellow spur and are arranged on an upright spike. The seed is black and flattened with a circular wing on the outer edge.

Seed production: Toadflax will grow well in cultivation, and harvesting and separation of the seed is best done by hand because otherwise the seed is difficult to separate from capsule and leaf fragments.

129. *Linum bienne*: Pale Flax (native)

Features: Pale Flax is a pale blue-flowered native biennial of meadows, pastures and grassy places mainly in the south of Britain and Ireland, and often near the sea. In flower it is similar to Linseed (*Linum usitatissimum*). Linseed, often naturalised in the wild as a relic of cultivation, is an annual and is not branched at the base. Pale Flax is branched at the base.

Seed production: Pale Flax is readily cultivated at a field scale and the seed may be harvested by combine harvester. It is a good restoration species for use within its native range.

Pale Flax (*Linum bienne*).

130. *Linum catharticum*: Fairy Flax (native)

Features: Fairy Flax is a widespread but declining winter annual of dry grasslands, meadows and pastures on chalk, limestone and sands.

Seed production: Fairy Flax is an important restoration species, but it is difficult to grow and harvest in cultivation. However, the seed is readily picked up when brush harvesting a meadow and may sometimes be cleaned out from mixtures. Fairy Flax establishes easily in restored and created meadows, and survives well under meadow management.

LEFT: Fairy Flax (*Linum catharticum*).

BELOW: Seed of Fairy Flax.

131. *Lotus corniculatus*: Bird's-foot Trefoil (native)

Features: Bird's-foot Trefoil (Bacon and Eggs) is a frequent and widespread cushion-forming perennial of well drained neutral to calcareous soils and is found in meadows, pasture, downland, rocky places, shingle and sand dunes. The clustered flowers are yellow and sometimes tinged in shades of dark orange. It is a highly variable species, and distinct local ecotypes are found across the range of occupied habitats (Mortimer, 2016; Smith et al., 2007). In flower, Bird's-foot Trefoil resembles Horseshoe Vetch. In seed, Bird's-foot Trefoil has linear pods grouped three to five resembling a bird's foot, and in this way differs from Horseshoe Vetch.

Seed production: Bird's-foot Trefoil is a very important restoration species. It grows well in cultivation, but will only combine directly following a period of sustained dry/warm

Bird's-foot Trefoil (*Lotus corniculatus*).

Seed pods of Bird's-foot Trefoil.

Seed of Bird's-foot Trefoil.

weather sufficient to dry the pods. A preferred method is to swathe the crop at maturity, allow it to dry in the swathe, and then pick it up with a combine fitted with a draper header. Bird's-foot Trefoil is a proscribed species and is subject to the Seed Marketing Regulations 2011. Regardless, wild native origin Bird's-foot Trefoil remains an important restoration species, and it is better to use wild native ecotypes rather than the proscribed robust forage cultivars.

132. *Lotus pedunculatus*: Greater Bird's-foot Trefoil (native)

Features: Greater Bird's-foot Trefoil is a widespread scrambling perennial of wetland found in wet meadows, marshy places and waterside habitats. It differs from Bird's-foot Trefoil in having hollow stems.

Seed production: The seed of Greater Bird's-foot Trefoil may be produced and harvested in a way similar to that for Bird's-foot Trefoil, but neither is an easy species.

133. *Lycopus europeus*: Gypsywort (native)

Features: Gypsywort is a rhizomatous perennial of wet places. The small flowers are white with purple dots and are borne in whorls in leaf axils.

Seed production: Gypsywort can be grown in cultivation using horticultural methods, with seeding inflorescences harvested by hand at maturity for drying and seed separation.

Gypsywort (*Lycopus europeus*).

134. *Lysimachia arvensis*: Scarlet Pimpernel (native)

Features: Scarlet Pimpernel (former Latin name *Anagallis arvensis*) is a widespread and important winter/spring annual of cultivated and disturbed ground, dunes, rocky places and sea cliffs. The flowers are usually scarlet, although sometimes, and rarely, they can be white, pink or blue. The petals are usually fringed with three-celled glandular scarlet hairs.

Scarlet Pimpernel (*Lysimachia arvensis*).

Seed of Scarlet Pimpernel.

The seed capsule is spherical and is slow to dehisce unless crushed, and then it splits into two halves, like an Easter egg.

Seed production: Scarlet Pimpernel is a low-growing, spreading plant, impossible to pick up with a combine harvester. However, it can be raked into swathes, allowed to dry, and then picked up with a combine harvester fitted with a draper header. Seed yield can be good.

135. *Lythrum salicaria*: Purple Loose-strife (native)

Features: Purple Loose-strife is a tall perennial frequent on the banks of slow-moving or still water, and in marshes and fens. The upright, tall flower spikes are densely packed with dark-pink flowers, and these flowers mature to produce pods containing large numbers of very small seeds.

Seed production: Purple Loose-strife may be grown on damp ground, and the inflorescences are best harvested complete by cutting and placing on a sheet to dry. As the inflorescences dry they release large quantities of very fine seed. Purple Loose-strife is a good restoration species for created wetlands because it is a widespread species and it establishes readily from seed.

Purple Loose-strife (*Lythrum salicaria*).

136. *Malva moschata*: Musk Mallow (native)

Features: Musk Mallow is a short-lived perennial of well-drained neutral to calcareous soils; it is distributed throughout most of the British Isles in grasslands and waysides, but is declining. The leaves are finely dissected, and the attractive mid-pink flowers are borne in bunches. The seeds are reasonably long lived and can form a persistent seed bank in the soil. It is a good restoration species.

Musk Mallow (*Malva moschata*).

The hairy seed of Musk Mallow.

Seed production: Musk Mallow establishes readily from direct precision drilling, and grows well in cultivation, doing best on calcareous soils. Conveniently for the combine harvester, the furry black kidney-shaped seed is retained loose within a persistent papery calyx. Seed yields are very good.

137. *Malva sylvestris*: Common Mallow (archaeophyte)

Features: Common Mallow is a very robust, large-seeded and common perennial of waysides and wasteland.

Seed production: Common Mallow grows well in cultivation at a field scale, producing large plants and good quantities of seed. The seed is retained on the plant and readily harvested with a combine harvester.

Common Mallow (*Malva sylvestris*).

138. *Matricaria chamomilla*: Scented Mayweed (archaeophyte)

Features: Scented Mayweed is a small, fragrant annual of cultivated fields on neutral to slightly alkaline soils that are not over-fertile. The whole plant is pleasantly scented when crushed, and the daisy-like flowers have a hollow conical receptacle, which distinguishes them from other similar species. Scented Mayweed has a long history of use in herbal medicine, being mentioned by Hippocrates (fifth century BC). It has been used to treat a wide range of conditions, administered as an oil extract or as an aqueous infusion (Otto et al., 2015).

Seed production: Scented Mayweed is grown commercially as a herb, and may also be used as a restoration species when creating arable weed communities. It establishes readily from precision drilling, and grows well in cultivation at a field scale. Seed harvesting can be by combine harvester with the drum set open and aspiration well down. Seed yield can be very good, and cleaning is not a problem.

Pineappleweed (*Matricaria discoidea*).

139. *Matricaria discoidea*: Pineappleweed (naturalised neophyte)

Features: Pineappleweed differs from Scented Mayweed in that it lacks ligulate ray florets. It is a common and widespread scented annual, first recorded in the wild as an escapee from Kew Gardens in 1869, and soon becoming one of the fastest spreading species in the British Isles.

Seed production: Pineappleweed is now a familiar species of paths, disturbed ground and cultivation, and seed is easily collected by hand wherever it is found.

140. *Medicago lupulina*: Black Medick (native)

Features: Black Medick is a common procumbent biennial, or winter annual, of neutral to calcareous soils, and may be found in meadows, pasture and grassy places. It has

Black Medick (*Medicago lupulina*).

Seed of Black Medick.

a persistent soil seed bank and will often spring from disturbed ground. In flower it is similar to Lesser Trefoil (*Trifolium dubium*), but in seed, the clusters of black curved seeds are unmistakable.

Seed production: Black Medick is easily cultivated. The seeds, at first green, turn black and are retained on the plant. Seed may be harvested directly by combine harvester, and seed yield is often heavy. Black Medick is a proscribed species subject to the Seed Marketing Regulations 2011. It is a forage species, with agricultural cultivars often sown for game cover. However, the wild native form of Black Medick is an important restoration species for meadows and pastures. It is quick to establish from restoration seeding, growing well in the second year after seeding, then giving way to species that establish more slowly.

141. *Mentha aquatica*: Water Mint (native)

Features: Water Mint is an uncommon rhizomatous perennial of very wet habitats close to open water such as ponds and rivers, wet woodlands, marshes and dune slacks. It is often partially or fully submerged, and may spread by detached rhizomes. Water Mint is widespread except in the Scottish Highlands

Seed production: Water Mint is a difficult species to grow in cultivation, and is best suited to small-scale horticultural propagation and hand collection of the seed.

Water Mint (*Mentha aquatica*).

Water Mint seed.

142. *Mercurialis perennis*: Dog's Mercury (native)

Features: Dog's Mercury is a widespread dioecious, clonal, rhizomatous perennial of woodland, hedgerow and shaded habitats. It is also found growing on north-facing slopes in the uplands and on limestone pavement in The Burren, the

Dog's Mercury (*Mercurialis perennis*).

only place it is found growing as a native in Ireland. Dog's Mercury has male and female flowers on separate plants, with female plants producing round grey seeds that resemble droplets of mercury.

Seed production: Dog's Mercury is an important woodland restoration species. It may be propagated from hand-collected seed or from root cuttings at a horticultural scale.

143. *Myosotis arvensis*: Field Forget-me-not (archaeophyte)

Features: Field Forget-me-not is a widespread winter annual, or biennial, of open habitats, particularly on disturbed or cultivated soils.

Seed production: Field Forget-me-not is a common weed of cultivation, and the black, shiny seed may be collected from wherever the plant may be growing. It may also be grown in cultivation at a field scale, with harvest by combine harvester. In all situations seed yield is only moderate.

144. *Neottia ovata*: Twayblade Orchid (native)

Features: Twayblade Orchid (former name *Listera ovata*) is a declining but widespread rhizomatous perennial of

Field Forget-me-not (*Myosotis arvensis*).

Twayblade Orchid (*Neottia ovata*).

calcareous to neutral soils; it may be found in woodland, hedgerows, meadows, waysides, quarries, dune slacks, heath and on limestone pavement. The green flowering spike arises from an opposite pair of stem leaves.

Seed production: Cultivation of Twayblade Orchid is difficult and seed production is very modest. Seeding spikes may be collected from plants in cultivation, but not from the wild. *The Biological Flora for Twayblade Orchid* has been published (Kotilinek et al., 2015).

145. *Odontites vernus*: Red Bartsia (native)

Features: Red Bartsia is a branched, hemi-parasitic annual herb of short, often over-grazed grassland, tracks and field edges, throughout most of Britain and Ireland. It is a variable species with mid- to dark-pink flowers that produce capsules with only a few seeds.

Seed production: Red Bartsia can sometimes be present in quantity as a weed of cultivation, and seed capsules may be hand collected or combine harvested. Seed yield is usually very poor, and the seed, which is small, white and furrowed, is light and very difficult to clean and separate from chaff.

Red Bartsia (*Odontites vernus*).

146. *Oenanthe pimpinelloides*: Corky-fruited Water-dropwort (native)

Features: Corky-fruited Water-dropwort is a winter-green tuberous perennial of hay meadows, pastures and waysides in southern England. It is probably not native to Ireland. The finely divided leaves form a winter rosette pressed to the ground, and this produces an erect summer inflorescence bearing tight umbels of white flowers. Corky-fruited Water-dropwort is generally considered not to be toxic, as far as is known, unlike many other species of the genus *Oenanthe*, but it is wise to be cautious.

Seed production: Corky-fruited Water-dropwort is a species that grows well in cultivation at a field scale, producing good quantities of seed. Mature seeds are retained on the inflorescence for several weeks and are very easily harvested by hand or combine harvester.

Corky-fruited Water-dropwort (*Oenanthe pimpinelloides*).

147. *Onobrychis viciifolia*: Sainfoin (native)

Features: Sainfoin is a perennial legume of calcareous grasslands. Robust agricultural cultivars have been grown for forage on farms since the seventeenth century, and many of the current wild populations may be relics and survivors of cultivation. Populations of smaller stature are found in the chalk grasslands of the south of England, and these populations may be native. The flowers are deep pink, streaked with red, and they each mature to produce a largish one-seeded pod that is reticulately ridged and toothed.

Seed production: Sainfoin is a proscribed species subject to the Seed Marketing Regulations 2011. Nevertheless, it is an important restoration species, and putative native wild stock is easily grown and harvested in cultivation at a field scale.

Sainfoin (*Onobrychis viciifolia*).

The sculptured seed of Sainfoin.

148. *Ononis repens*: Common Restharrow (native)

Features: Common Restharrow is a procumbent, rhizomatous perennial legume of calcareous and sandy soils inland, and on sand dunes and stable shingle by the coast. It is present in the wild as a common spine-free morpho-type, and as an uncommon spiny morpho-type, as is evident when touching the plant. Spiny Restharrow (*Ononis spinosa*) is a similar, but different species. It is an ascending perennial sub-shrub, and is always spiny.

Seed production: Common Restharrow is not easy to grow in cultivation and produces only small amounts of seed as the species relies, partially and importantly, on vegetative rhizomatous spread.

Common Restharrow (*Ononis repens*).

149. *Ophioglossum vulgatum*: Adder's Tongue (native)

Features: Adder's Tongue is a rhizomatous deciduous fern of mildly acidic to calcareous soils. It is increasingly rare, inconspicuous, and a delight when spotted in spring. It may be found in meadows, damp pasture, woodland rides, sand dunes, fen and heathland. The leaf is a simple entire blade, from the base of which arises a single fertile spike with sunken sporangia.

Spore production: Adder's Tongue and the related Moonwort (*Botrychium lunaria*) are potentially important and delightful restoration species; however, both are hugely challenging to grow in cultivation and obtain spores from.

Adder's tongue (*Ophioglossum vulgatum*).

150. *Ophrys apifera*: Bee Orchid (native)

Features: Bee Orchid is an uncommon mono-carpic, wintergreen, self-fertilising, tuberous perennial of neutral to calcareous soils, and is found occurring across a wide range of habitats. Originally restricted to the south and east of England (Keble Martin, 1965) its range has increased dramatically since the 1960s, and it appears to be migrating north – it has even been found at several sites north of the southern Highlands in Scotland.

Although considered to be monocarpic, and most often it is, Wells and Cox (1991), through careful study, observed that some plants may flower in up to three successive years. The normal life cycle of a Bee Orchid, from seed germination, is three to four years underground partnered with a Bee Orchid specific mycorrhizal fungus, and three to four years above ground as wintergreen rosettes; then in years six to eight, individual plants flower and die.

Pollination of the flowers is crucial for survival of the species. Bee Orchid is a bee mimic, seeking to attract specific male bees by emitting allomone attractants, and by having a female bee-like labellum. In continental Europe Bee Orchid is frequent and is often cross-pollinated by pseudo-copulation visits from male bees. In Britain and Ireland these specific pollinating bee species are absent or rare, and so cross-pollination is rare.

In the absence of cross-pollination, the filaments supporting the pollinia curve downwards, placing the pollinia on the stigma of its own flower, and often the filaments will cross in doing so. This enables and ensures self-fertilization as a fall-back option, and this is critically important as Bee Orchids are mostly monocarpic. Variation in the form of the Bee Orchid flower can be considerable within populations. This variation is consistent with occasional cross-pollination events followed by self-pollination of the F1, the observed segregation of flower features occurring especially in the S1 generation.

The following images show a variation in flower characters within a single large population of Bee Orchid.

Bee Orchid with pale sepals and upturned edge to the labellum.

Actinomorphic Bee Orchid with petals reduced and ± equal.

Bee Orchid with reduced upper petals and obscurely marked labellum.

Bee Orchid with long narrow upper petals.

Bee Orchid with pale sepals and large pale upper petals.

Bee Orchid with greenish upper petals and obscurely marked labellum.

Bee Orchid with large pink upper petals and small side lobes to the labellum.

Seed production: The pollinated flowers of Bee Orchid produce large capsules containing dust seed. Spikes bearing capsules may be collected by hand at the late yellow pod stage, but only from plants in cultivation, and not from the wild. The capsules may be crushed when dry to release the dust seed, taking care to do so where there is no air movement.

151. *Orchis mascula*: Early-purple Orchid (native)

Features: Early-purple Orchid is a polycarpic tuberous perennial of neutral to calcareous soils throughout most of Britain and Ireland. Its range is declining, but it may be found in woodland, scrub, meadows, pasture, waysides and coastal grasslands. The wintergreen leaves are often dark spotted, and the flowers bloom in spring, varying in shade from pale pink to the more frequent dark pink; they smell unpleasantly of cats.

Seed production: Seedpods can be collected from plants in cultivation, or from waysides or meadows that are due to be mown. Seed, or any other plant material, must not be collected from the wild.

Early-purple Orchid (*Orchis mascula*) with unspotted leaves.

152. *Origanum vulgare*: Wild Marjoram (native)

Features: Wild Marjoram is a wiry-stemmed, aromatic, long-lived perennial of calcareous soils, where it may be widespread and locally common. It is an attractive, late-flowering species of grasslands, waysides and hedgerows, and is tolerant of poor soils and drought. It is a good restoration species, being excellent at attracting butterflies and bees, and in providing pollinator services. It is avoided by most grazing livestock.

Seed production: Wild Marjoram germinates well from an autumn or spring drilling in the field. The seedlings are at first very small, and grow slowly in cultivation, producing mature, seed-bearing plants in the second or third year after sowing. The seed is very small, round and brown in colour. Judging readiness to harvest is difficult as the seed is

Wild Marjoram (*Origanum vulgare*).

Seed of Wild Marjoram.

very inconspicuous. As a rough guide, test combine with the aspiration low and when inflorescences start to turn from deep brown to silver edged. The swathe may be picked up again a few days after the first harvest, and this will yield an additional moderate amount of seed. Overall yield will be, at best, around 20kg/acre, and the crop annoyingly repeatedly blocks the straw walkers.

153. *Orobanche minor*: Common Broomrape (native)

Features: Common Broomrape is a handsome, non-green, obligate root parasite on a wide range of host species. It is an annual producing an unbranched, leafless inflorescence spike, with flowers coloured yellow and tinted purple. It is a variable and evolving species with many different ecotypes and forms. Other species in the genus *Orobanche* are less common and often quite restricted in host range.

Seed production: Common Broomrape can sometimes be very abundant in restored meadows, providing an opportunity to gather mature seeding spikes by hand. The seed is dark brown and dust-like.

Common Broomrape (*Orobanche minor*).

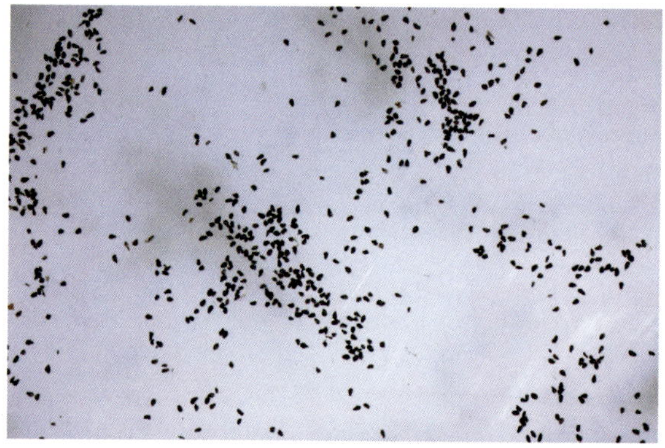

The minute seed of Common Broomrape.

154. *Papaver dubium*: Long-headed Poppy (archaeophyte)

Features: Long-headed Poppy is a widely distributed winter annual of cultivated ground on calcareous to neutral soils, with red flowers similar to those of Common Poppy.

The seedpod of Long-headed Poppy (*Papaver dubium*).

It is distinguished by having elongated seed capsules and slightly larger seeds compared to the roundish seed capsules and small seeds of Common Poppy. The Yellow-juiced Poppy (*Papaver lecoqii*) differs from Long-headed Poppy in being mostly restricted to heavy calcareous soils, and in bleeding yellow juice from the stem when snapped. Long-headed Poppy bleeds white juice, as does Common Poppy.

Seed production: Long-headed Poppy often grows with Corn Poppy, and it is rare to find Long-headed Poppy as an agricultural crop contaminant. However, it may be grown alone as a crop in the same way as Corn Poppy is grown and harvested.

155. *Papaver rhoeas*: Corn Poppy (archaeophyte)

Features: Corn Poppy is an abundant and widespread winter annual of cultivated and disturbed soils throughout Britain and Ireland, and is an important restoration species. It produces abundant fine seed that is very long lived. The seed has no dormancy requirements, and has a half-life of forty years buried in the soil. Corn Poppy is likely to spring from the soil seed bank wherever soil is cultivated, exposing buried seed to light. Most agricultural grassland (in fields that are not bumpy) will have been ploughed and harrowed in the past for arable cropping, and most of these fields will carry a buried soil seed bank of viable Corn Poppy seed. Corn Poppy is only prevented from dominating arable landscapes by the liberal application of herbicides. Where it does escape the sprayer, poppies can bloom in the run-up to cereal harvest. The four large red petals are at first crinkled from tight packing when in bud.

Common Poppy (*Papaver rhoeas*).

Seed of Common Poppy.

Seed production: Cleanings from unsprayed winter cereals are a good source of Corn Poppy seed. It may also be grown alone as a crop. The fine seed may be precision drilled in rows in the late summer/early autumn, and germination will occur before the onset of winter. The plants will grow through the winter and the following year, to reach seed maturity in July. They are well presented for harvest, with upright inflorescences and the fine seed held loose in pepperpot-like capsules. Set the concave closed to break the capsules and the aspiration down to retain the seed.

156. *Paris quadrifolia*: Herb Paris (native)

Features: Herb Paris is a distinctive rhizomatous perennial of deciduous woods. It produces flowers in the centre of a whorl of four leaves, and these flowers produce a small cluster of black berries.

Seed production: Cultivation and reproduction by seed is not easy. This species is better suited to vegetative propagation.

Herb Paris (*Paris quadrifolia*).

157. *Pastinaca sativa*: Wild Parsnip (native)

Features: Wild Parsnip is a hollow-stemmed, wintergreen biennial umbellifer of waysides and grassy places on calcareous soils in England and Wales. The plant does not have a swollen root, but it has a characteristic parsnip aroma when crushed. The flowers are yellow/green in colour and produce large flat seeds that require vernalisation to break dormancy.

Seed production: Wild Parsnip is readily cultivated at a field scale and the seeds may be combine harvested. It is best autumn drilled, and will germinate in the following spring and grow vigorously. The first harvest will be in the year after.

Wild Parsnip (*Pastinaca sativa*) and the Hogweed Bonking Beetle (*Rhagonycha fulva*).

Seed of Wild Parsnip.

158. *Phleum arenarium*: Sand Cat's-tail (native)

Features: Sand Cat's-tail is a small annual grass frequent on coastal sand dunes and vegetated shingle, and inland in East Anglia on sandy heaths and disturbed sandy areas.

Seed production: Sand Cat's-tail is a diminutive, non-competitive grass, best grown in short rows and harvested by gathering the small seed heads by hand.

159. *Phleum bertolonii*: Smaller Cat's-tail (native)

Features: Smaller Cat's-tail is a common and widespread perennial grass of meadows, pastures and other grassy places on neutral to calcareous soils. It is a variable species, varying considerably in stature from place to place, but it is always smaller than, and rather earlier flowering than Timothy *(Phleum pratense)*, a species grown on farms for fodder.

Small Timothy (*Phleum bertolonii*) at anthesis.

Seed of Small Timothy.

Seed of Mouse-ear Hawkweed.

Seed production: Smaller Cat's-tail is a proscribed species and is subject to the Seed Marketing Regulations 2011. However, it is an important restoration species and wild stock is easy to grow and harvest in cultivation at a field scale.

160. *Pilosella officinarum*: Mouse-ear Hawkweed (native)

Features: Mouse-ear Hawkweed is a low-growing, mat-forming stoloniferous perennial of dry, infertile calcareous, neutral and acidic soils. It is common and widespread in short grassland and on sand dunes, cliffs and rocky places. The stolons are clothed in simple leaves that are softly hairy and distinctly white on the underside. The inflorescence is unbranched and variously clothed in simple, branched and glandular hairs. The small Dandelion-like flowers are lemon yellow in colour, and mature to produce a head of tiny ribbed black seeds with a whitish pappus and no beak.

Mouse-ear Hawkweed (*Pilosella officinarum*).

Seed production: Mouse-ear Hawkweed is a demanding species to grow in cultivation, and seed harvest, which is usually by hand, requires still weather. The pappus is easily removed by rubbing.

161. *Pimpinella major*: Greater Burnet-saxifrage (native)

Features: Greater Burnet-saxifrage is a large-leaved, hollow-stemmed, perennial umbellifer of waysides, woodland borders and hedgerows on calcareous soils. It is locally common in Britain and Ireland, but rather rare in Wales and Scotland.

Seed production: Greater Burnet-saxifrage is easily grown in cultivation for seed. It is best autumn drilled in the field, and will germinate in the following spring. The first harvest, which can be combine harvested, will be in the year after.

162. *Pimpinella saxifraga*: Burnet-saxifrage (native)

Features: Burnet-saxifrage is a late-flowering, white-flowered, solid-stemmed, perennial umbellifer of neutral to

Burnet-saxifrage (*Pimpinella saxifraga*).

calcareous, well-drained soils. It has finely divided leaves and is common and widespread on waysides, rocky places and grasslands, especially in chalk and limestone regions.

Seed production: Burnet-saxifrage can be grown in cultivation, and well-grown plants can yield good quantities of seed. The seed may be harvested easily by a variety of methods; when placed in a sack, it can slowly release an unpleasant odour.

163. *Plantago coronopus*: Buck's-horn Plantain (native)

Features: Buck's-horn Plantain is a rosette-forming, short-lived perennial halophyte, common in coastal locations and increasingly spreading inland along salt-treated roads. The leaves are deeply and narrowly lobed, and the plant is tolerant of dry, infertile soils and trampling.

Seed production: Seed may be collected by hand from plants in cultivation as the flowering spike matures. Small amounts may be collected from abundant populations elsewhere, provided it is permitted to do so.

164. *Plantago lanceolata*: Ribwort Plantain (native)

Features: Ribwort Plantain (Fire Weed) is a widespread and common rosette-forming wintergreen and short-lived perennial of calcareous, neutral and slightly acidic soils. It is frequently abundant in meadows, pastures, waysides, wasteland and sand dunes, and in lawns and amenity grasslands. It has ribbed, strap-like leaves and is tolerant of grazing, mowing, trampling and a wide range of soil conditions. Plantain leaves are very slow to dry in mown hay, and if baled too soon and carted to the barn, patches of Plantain in the bales can heat and burn the barn down.

Ribwort Plantain (*Plantago lanceolata*) at anthesis.

Seed of Ribwort Plantain.

Seed production: Ribwort Plantain grows exceedingly well in field cultivation, reaching seeding maturity within twelve to eighteen months from drilling. However, the crop may only persist for a year or so before stem rot (*Rhizoctonia?*) reduces yield. A healthy crop will produce a heavy yield, harvested by combine harvester with the drum closed and the aspiration up. The shiny brown, torpedo-shaped seed has no dormancy, good longevity, and the seeds are myxospermous, having an outer testa that swells when wet to form a coat of sticky mucilage (Cowley and Burton, 2020). This mucilage may aid dispersal and germination, but if harvested seed gets wet it can set like concrete on drying. Ribwort Plantain seeds glue to each other and to most things, but they do not glue to the few seeded transversely dehiscing capsules from which they originate.

165. *Plantago media*: Hoary Plantain (native)

Features: Hoary Plantain is a common, wintergreen, rosette-forming perennial of calcareous soils over chalk or limestone. It is a frequent species of meadows, pasture, waysides and short grasslands on calcareous substrates. The leaves are ribbed, rounded, finely downy and appressed to the ground. The inflorescence spike is unbranched and erect, with flowers arranged so that anthesis progresses from the bottom up, rather like a 'sparkler' in reverse.

Seed production: Seed may be directly precision drilled in rows in spring or autumn. Plants grow well, and at seed maturity a combine harvester may harvest the crop, with the drum closed and the aspiration moderate. Seed yield is usually good, although less than that of Ribwort Plantain.

166. *Platanthera chlorantha*: Greater Butterfly-orchid (native)

Features: Greater Butterfly-orchid is a tuberous perennial of open to partially shaded situations on calcareous, neutral, or sometimes slightly acidic soils. It is more frequent in the south, and can occasionally form population clusters in late-cut meadows, pasture and open woodland. The Lesser Butterfly-orchid (*Platanthera bifolia*) is more frequent in the north, occurring in similar habitats. The

Hoary Plantain (*Plantago media*) at anthesis.

Greater Butterfly-orchid (*Platanthera chlorantha*).

Greater Butterfly-orchid has a pair of rounded, unmarked basal leaves and produces a handsome spike bearing long, spurred, greenish-white flowers. The flowers have a recurved strap-like labellum and a pair of pollinia positioned to attach to visiting pollinators. The spur has a nectar droplet at its far end, and the flowers release a moth-attracting scent at night. Visiting moths pollinate the flowers.

Seed production: Seed capsules may be collected from plants in cultivation and slowly dried to release dust seed that is the colour of ground cocoa. Seed capsules should not be collected from the wild, and the species should not be sown outside its native range.

167. *Polygala vulgaris*: Common Milkwort (native)

Features: Common Milkwort is a small, procumbent wintergreen perennial of calcareous, neutral and slightly acid soils. It is a declining species of meadows, pastures, heathland, sand dunes and rocky places, with populations scattered throughout the British Isles. The small leaves are simple and alternate, and the zygomorphic flowers are often present in shades of blue and pink.

Seed production: Common Milkwort is a small, non-robust species, and cultivation is best done using the application of horticultural methods rather than field-scale agricultural methods.

Common Milkwort (*Polygala vulgaris*).

168. *Polypodium vulgare*: Polypody (native)

Features: Polypody is a wintergreen, rhizomatous, long-lived perennial fern, frequent in the west. It is a common fern on walls, banks, rocky places, scree, sand dunes, and as an epiphyte in deciduous woodland, especially on Oak trees. It is tolerant of exposure and can recover from desiccation. The fronds are once pinnate with parallel-sided, rounded pinnae, and the sori on the undersides of the fronds are usually orbicular in shape.

Polypody (*Polypodium vulgare*).

Spore collection: Sporulating fronds of Polypody may be collected whole, in moderation and where the plant is abundant and it is permitted to do so. These fronds may be transferred directly to the restoration site or may be dried to release the fine, dust-like spores.

169. *Poterium sanguisorba*: Salad Burnet (native)

FEATURES: Salad Burnet (former Latin name *Sanguisorba minor)* is a very deep-rooted, long-lived perennial of calcareous soils. It is frequent in meadows and pasture, and is a characteristic species of dry and infertile grasslands on

Salad Burnet (*Poterium sanguisorba*) at anthesis.

Seed of Salad Burnet.

chalk and limestone. The leaves are once pinnate, and the pinnae are toothed. The flowers are borne in terminal clusters, usually with the female flowers uppermost (flowering first), and male flowers lowermost (flowering later). Salad Burnet is almost exclusively wind pollinated and is highly self-compatible. It can set seed by autogamous selfing in the absence of cross-pollination (Abrahamczyk et al., 2023). The angular, buff-coloured seed has a reticulate surface, but the ridges on the angles are not winged. The related non-native Fodder Burnet (*Poterium sanguisorba ssp balearicum*) has winged angles to the seed, and the plants are usually more robust.

Seed production: Cultivation and seed harvest of native Salad Burnet is not difficult, but seed yield is modest compared to that of Fodder Burnet.

170. *Primula veris*: Cowslip (native)

Features: Cowslip is a frequent and widespread rosette-forming diploid (2n=2x=22) perennial of well-drained calcareous and neutral soils. It is a spring-flowering species of meadows, pastures, waysides, hedgerows, downland and other grasslands, and is a variable species with local ecotypes. The puckered leaves have a characteristic

Cowslip (*Primula veris*) and pollinating bee fly (*Bombylius major*).

Slowly maturing seed capsules of Cowslip.

Seed of Cowslip.

Primula scent, and the sweet yellow flowers are held in bunches on an upright short peduncle. Plants are distylous and heterostylous and are represented by two flower morphs, either 'pin' or 'thrum', a floral structural arrangement that ensures that pollinators cross-pollinate between pin and thrum plants but do not self-pollinate individual Cowslip plants, and do not cross-pollinate within pin or thrum plant populations.

Cowslip is an obligate out-breeder, and the flowers partially droop when attracting pollinator insects. After pollination (during May), the pedicels straighten to produce an upright bunch of maturing capsules. The calyx tube is persistent, and the seed-containing capsule, within the calyx, very slowly matures over a period of six to eight weeks. As seed maturity nears in late June/early July, the seeds turn from green to velvet brown, the apex of the seed capsule extends into a rounded translucent cone, and the cone splits into a toothed, crown-like opening. At this point the gelatinous receptacle to which the seeds are attached, gradually dries, and the brown seeds become loose within the capsule.

Seed production: Cowslip is an important restoration species, responding well to hay-meadow management (Brys et al., 2004), and it is easily cultivated. The seeds require vernalisation to break dormancy and will germinate in the spring from an autumn drilling, or will germinate in the following spring from a spring drilling. Well-grown mature plants can flower profusely, and with good pollination seed yield can be excellent. The seeds are retained loose in the upright open capsules and may be harvested at leisure by hand or combine harvester. *The Biological Flora for Cowslip* has been published (Brys and Jacquemyn, 2009).

171. *Primula vulgaris*: Primrose (native)

Features: Primrose is a long-lived, rosette-forming diploid (2n=2x=22) perennial of shady places on neutral to calcareous soils and is widely distributed and frequent in woodland, scrub, hedgerows and on waysides and shady banks. The yellow flowers, opening in early spring, are borne individually facing up, and pollination is reliant on early emerging bumblebees. Seed set varies from year to

Primrose (*Primula vulgaris*).

Seed of Primrose.

year dependent on weather, and in a cold, windy spring seed set can be very poor. Primrose, like Cowslip, is self-incompatible and has pin and thrum flower forms. Cross-pollination must occur otherwise no seed is set. Each fully and successfully pollinated flower will produce around fifty seeds, and these take six to eight weeks to mature in the swelling capsule.

Shortly after pollination the pedicel coils downwards, placing the fading flowers under the puckered leaves. Seeds turn from green to brown in mid-June, and the seed capsule swells at the tip and opens like a crown. The seeds of Primrose have a nutritious external elaiosome (the seeds of Cowslip do not have an elaiosome), and ants and small beetles out foraging for seed soon visit the just-opened pods.

Seed production: Primrose is an important restoration species and grows well in cultivation. The seed requires vernalisation to break dormancy, and seed may be precision drilled in the field in the autumn for spring germination. Seedpod harvesting is by hand at the brown seed stage just before the capsules' dehiscence, and pods are collected from beneath the rosettes by breaking the pedicels. Seed yield can be very poor if pollination is poor and if invertebrates have robbed the seed. The seedpods need to be dried slowly to allow for complete maturation of the seed and for the gelatinous receptacle to shrink. The seeds will fall loose from the pods as the receptacle dries.

172. *Prunella vulgaris*: Selfheal (native)

Features: Selfheal is a common and widespread short-lived perennial of calcareous, neutral and acid soils. It may be found in meadows, pastures, waysides, lawns and amenity grassland. The leaves are entire and slightly glossy, and the flowers are bluish violet in colour.

Seed production: Selfheal is an easy restoration species, growing well in cultivation and producing large quantities of seed from a maiden and first main harvest. However, Selfheal is susceptible to stem rot (*Rhizoctonia?*) in cultivation, and cropping may be short lived. The light-brown seed is mucilaginous (myxospermous) and will germinate readily without dormancy restrictions.

Selfheal (*Prunella vulgaris*).

Seed of Selfheal.

173. *Pulicaria dysenterica*: Common Fleabane (native)

Features: Common Fleabane is a densely hairy, extensively rhizomatous, erect perennial of marshes, ditches and wet fields. It is common in the south and rare in Scotland. Common Fleabane is late flowering, producing rather attractive yellow capitula with neatly arranged ray and disc florets.

Seed production: Common Fleabane may be cultivated, and fading flowers may be collected by hand. After drying, the seed can be separated from the stalks and the light-brown pappus removed by rubbing.

Common Fleabane (*Pulicaria dysenterica*).

Seed of Common Fleabane.

174. *Ranunculus acris*: Meadow Buttercup (native)

Features: Meadow Buttercup is a very common and widespread perennial of meadows, pasture, waysides and other grassy places on calcareous, neutral and slightly acid soils. It is an erect plant and differs from Bulbous Buttercup (*Ranunculus bulbosus*) in having clasping sepals at anthesis. Buttercup species release irritant volatiles when crushed or eaten, and are generally avoided by grazing cattle and equines. However, these volatiles disperse during haymaking, and dry hay containing Buttercups is harmless and palatable.

Seed production: Meadow Buttercup is an important mid-season restoration species for meadows, and seed crops may be produced using standard cultivation and harvesting methods. The seeds require vernalisation to

Meadow Buttercup (*Ranunculus acris*) with a clasping calyx.

Seed of Meadow Buttercup.

break dormancy so are best precision drilled in the autumn for spring germination. The seeds are smooth and dark brown to black when mature, and are loosely attached to the inflorescence. Harvest is best by combine harvester, but timing is critical as the seeds are easily dislodged by wind.

175. *Ranunculus arvensis*: Corn Buttercup (archaeophyte)

Features: Corn Buttercup is a very rare annual of cultivated fields; it was formerly more widespread.

Seed production: Corn Buttercup may be grown in cultivation for seed, which is harvested by hand or combine harvester. The seeds are quite large and are covered with conspicuous spines. Sometimes Corn Buttercup is harvested as a crop contaminant, but the seeds tangle with other contaminants and are very difficult to clean out.

Seed of Corn Buttercup (*Ranunculus arvensis*).

176. *Ranunculus auricomus*: Goldilocks (native)

Features: Goldilocks is an often overlooked Buttercup of woods and thickets and other places on calcareous soils. It is widespread, and frequent and similar to Meadow Buttercup except that the basal leaves are reniform in shape and the achenes are hairy. Goldilocks is an interesting and complex species, consisting of different ploidy levels, some of which are apomictic and some reproduce sexually (Barke et al., 2018).

Seed production: Goldilocks may be cultivated for seed in a similar way to other species of *Ranunculus*.

Goldilocks (*Ranunculus auricomus*).

177. *Ranunculus bulbosus*: Bulbous Buttercup (native)

Features: Bulbous Buttercup is a common and widespread perennial of meadows, pastures and other grasslands. It flowers a little before Meadow Buttercup, and has a preference for better-drained soils. Bulbous Buttercup is distinguished by having strongly reflexed sepals at anthesis, a swollen bulb-like stem base, and spreading inflorescences angled at around 45 degrees.

Seed production: Bulbous Buttercup is an important restoration species for meadows and may be grown in cultivation for seed. The seed is best precision drilled in the autumn for spring germination. Sometimes Bulbous Buttercup is sufficiently abundant in a meadow to brush harvest for seed in mid-June. At this time of the season few other meadow species are shedding seed, and with careful cleaning, Bulbous Buttercup may be separated from the rest. It has seed larger than that of Meadow Buttercup and it is lighter in colour.

Bulbous Buttercup (*Ranunculus bulbosus*) with a reflexed calyx.

Seed of Bulbous Buttercup.

178. *Reseda lutea*: Mignonette (native)

Features: Mignonette is a widespread biennial, sometimes short-lived perennial of calcareous soils. It is a species of open ground, disturbed soils, chalk and limestone grasslands, quarries and rocky places. The erect inflorescent spikes are yellowish in flower, and as these fade, black shiny round seeds form in papery capsules.

Seed production: Mignonette is a species that may be precision drilled on calcareous ground where it will grow well. The crop can be combine harvested, producing good quantities of seed.

Mignonette (*Reseda lutea*).

179. *Rhinanthus minor*: Yellow Rattle (native)

Features: Yellow Rattle is common and widespread, and is an obligate hemiparasitic spring annual. It is parasitic on a range of grassland species and can be frequent in meadows, pastures and waysides on a wide range of soils in both the

Yellow Rattle (*Rhinanthus minor*).

Seedlings of Yellow Rattle with characteristic crinkled first true leaves.

lowlands and the uplands. It is a variable species, and six subspecies are recognised in Britain and Ireland. These appear to be ecotypes, but the variation between these subspecies may in fact be due to the effect of the chosen

Seed of Yellow Rattle.

host on the morphology of the parasitising Yellow Rattle. Houston and Wolff (2012) considered there to be insufficient genetic evidence for the existence of ecotypes of Yellow Rattle in the UK, and therefore there was no advantage in using local origin seed of this species, but rightly maintained that native British and Irish origin stock remained important for restoration projects.

Seed production: Yellow Rattle is an exceedingly important restoration species because it can suppress competitive grassland species and can enable the restoration and establishment of species-diverse meadows. Yellow Rattle seed is normally obtained by brush harvesting Rattle-rich hay meadows in early July. The harvested mixture may then be cleaned to separate out the large flat seeds of Yellow Rattle. *The Biological Flora for Yellow Rattle* has been published (Westbury, 2004).

180. *Rubia perregrina*: Madder (native)

Features: Madder is a scrambling evergreen perennial of mainly coastal locations on waysides, hedgerows, scrub and rocky places.

Seed production: The small, one-seeded succulent fruits of Madder may be collected by hand from the wild where plants are abundant and where it is permitted to do so.

181. *Rumex acetosa*: Sorrel (native)

Features: Sorrel is a short-lived perennial frequent in meadows, pastures and other grassy places on calcareous to neutral soils. In contrast to weedy species of the genus, Sorrel is nutritious and palatable to livestock, the leaves having a tangy flavour. It is a variable species, and the inflorescence,

Seed of Sheep's Sorrel (*Rumex acetosella*).

LEFT: Sorrel (*Rumex acetosa*).

BELOW: Seed of Sorrel.

sometimes tall and standing above other meadow species, is a rusty red and produces shiny angular seeds.

Seed production: In cultivation, seed crops of Sorrel establish easily from drilled seed and grow well to produce inflorescences that are well presented for combine harvesting.

182. *Rumex acetosella*: Sheep's Sorrel (native)

Features: Sheep's Sorrel is a low-growing rhizomatous perennial of impoverished acid soils and is common on dry heaths, dunes and non-calcareous rocky places.

Seed production: Sheep's Sorrel is not easy to grow in cultivation because of the requirement for an acid substrate, and seed harvest by combine harvester is not easy because of the low stature of the plants. In wild locations, Sheep's Sorrel can be abundant and form the dominant ground cover. Brush harvesting is often possible, as is hand harvesting.

183. *Salvia verbenaca*: Wild Clary (native)

Features: Wild Clary is a discretely attractive, wintergreen, wayside perennial, not to be confused with its showy cousin, Meadow Clary (*Salvia pratensis*), a protected and very rare

Wild Clary (*Salvia verbenaca*).

native of the British Isles. Wild Clary is frequent in hedgerows and unmanaged grassland on a variety of soils.

Seed production: Wild Clary grows easily in cultivation and produces seed that may be harvested by hand or combine harvester.

184. *Sanguisorba officinalis*: Great Burnet (native)

Features: Great Burnet is a clump-forming, long-lived perennial of floodplain meadows, wet meadows and other damp grasslands. It is widespread and locally frequent, but declining, and is later flowering than most meadow species. The leaves are once pinnate, and the flowers are held above the foliage in small, dense clusters. These are the colour of red wine and are ambophilous, being largely insect pollinated, but also sometimes wind pollinated (Abrahamczyk et al., 2023).

Seed production: Great Burnet may be grown in cultivation on damp ground, and seed clusters can be hand picked at maturity, or combine harvested. Seed yield is low, and plants in cultivation are susceptible to powdery mildew.

185. *Sanicula europeus*: Sanicle (native)

Features: Sanicle is a perennial umbellifer of woodland and shady places on calcareous soils. It forms a basal rosette of palmately lobed leaves, and has a leafless, branched inflorescence bearing small clusters of white flowers.

Seed production: Sanicle seeds are green and clothed in soft-hooked spines, and they may be collected by hand where plants are abundant.

ABOVE LEFT: Great Burnet (*Sanguisorba officinalis*).

ABOVE RIGHT: Hoverfly visiting a Great Burnet flower.

Seed of Great Burnet.

Sanicle (*Sanicula europeus*).

186. *Saxifraga granulata*: Meadow Saxifrage (native)

Features: Meadow Saxifrage is a rhizomatous perennial, forming a basal rosette of lobed leaves and axillary bulbils at the soil surface (Stroh, 2015c). It is not common, but locally frequent in meadows and pasture on calcareous soils. It has pretty white flowers that mature to produce a capsule containing black, dust-like seed.

Seed production: With care, Meadow Saxifrage may be grown in cultivation and the seed capsules collected by hand, with the seed being released on drying.

Meadow Saxifrage (*Saxifraga granulata*).

RIGHT: Small Scabious (*Scabiosa columbaria*).

BELOW: Seed of Small Scabious.

187. *Scabiosa columbaria*: Small Scabious (native)

Features: Small Scabious is a rosette-forming, long-lived perennial of dry calcareous grassland and rocky places. The clusters of blue flowers attract clouds of butterflies and many bees.

Seed production: Small Scabious may be precision drilled in rows, and establishes readily on calcareous ground. When mature it has an extended flowering period from mid- to late season; seed maturation is rather indeterminate. Best seed yields are obtained by hand plucking seed clusters just before individual seeds detach, and repeating this every few days through the season. Small Scabious can be combine harvested at peak maturity, but the seed yield obtained in this way is modest.

188. *Scrophularia auriculata*: Water Betony (native)

Features: Water Betony (Water Figwort) is a tall waterside and wetland perennial. It is common in suitable habitat in England and Wales, and appears to be extending its range further west and north in recent times.

Seed production: Water Betony may be grown in cultivation on wet ground, with seed capsules harvested by hand. On drying, the capsules release fine black seed.

189. *Sedum acre*: Biting Stonecrop (native)

Features: Biting Stonecrop is a cushion-forming perennial of nutrient-poor, calcareous, neutral and acidic substrates. It is grazing tolerant, and is frequent on walls, sand dunes, railway tracks, abandoned hard surfaces and rocky places.

Seed production: Biting Stonecrop is used occasionally for green roofs, as a wild native alternative to garden cultivars of *Sedum*, but it could be used far more, as it is a native species supporting native pollinators. It is a species that can be propagated vegetatively by standard horticultural methods, and seed may be collected by hand from mature plants.

Biting Stonecrop (*Sedum acre*).

190. *Serratula tinctoria*: Saw-wort (native)

Features: Saw-wort is a clump-forming perennial, scattered in England and Wales, and rare in Ireland and Scotland. It is found on calcareous, neutral and acid soils in meadows, pastures, heaths, cliff tops and waysides, but is declining in all places. The leaves are serrated, and the erect inflorescence is much branched to produce loose groups of purple flowers. Flowering is in mid-summer, and the seed heads mature to produce clusters of elongated grey seeds with a brown unbeaked pappus. The seeds remain attached to the capitulum for an extended period, and are enclosed by involucre bracts. Separation of the seeds and removal of the pappus is difficult, and germination of sown seed is often very poor.

Seed production: Saw-wort is an important restoration species, but very challenging to establish in cultivation. Seed yields are low, and seed harvest is never easy. *The Biological Flora for Saw-wort* has been published (Jefferson and Walker, 2017).

Saw-wort (*Serratula tinctoria*).

191. *Sherardia arvensis*: Field Madder (native)

Features: Field Madder is a declining annual of cultivated and disturbed ground. It is a variable clump-forming species with pointed, simple leaflets in stem whorls and pink flowers. Individual flowers mature to produce a pair of nutlets with a persistent calyx, both retained within the axil of a leaflet.

Seed production: Field Madder can be abundant on occasions, and mature plants may be harvested by hand. On drying, the seed is difficult to detach from the stem and leaf material, and seed yield can be disappointing.

Field Madder (*Sherardia arvensis*).

192. *Silaum silaus*: Pepper Saxifrage (native)

Features: Pepper Saxifrage is a declining, short-lived perennial umbellifer of neutral and calcareous grasslands. It has a southern distribution and is a notable species of floodplain meadows and neutral hay meadows. The finely divided leaves, one to four pinnate, smell of pepper when crushed, and the upright yellow/green inflorescences produce clusters of plump green seeds that dry, when mature, to ribbed brown seeds.

Seed production: This is a good species to grow in cultivation, germinating in the spring from an autumn drilling. Seed production in the following year is often very good,

Seed of Pepper Saxifrage.

and seed may be harvested by combine harvester, producing high yields.

193. *Silene dioica*: Red Campion (native)

Features: Red Campion is a common and widespread short-lived perennial of open woodland, waysides, scrub, hedgerows, coastal cliffs and rocky upland habitats. It is a

Red Campion (*Silene dioica*).

Seed of Red Campion.

variable, soft-textured, upright plant with pink to reddish-pink flowers. Red Campion is dioecious, with male and female flowers on separate plants, thus ensuring cross-pollination. Female plants are large flowered, and after cross-pollination produce capsule-containing seed that at maturity is reddish brown. Male plants are smaller, and smaller flowered, and produce no seed. The ratio of female to male is usually 1:1.

Seed production: The seed of Red Campion will germinate without dormancy restrictions, and direct drilling in autumn or spring will soon produce neat rows. These rows grow luxuriantly, and flower in the first full year after drilling. The crop is upright, and the female seedpods are held upright, dehiscing by a small aperture at the apex. The brown/red seed is retained loose within the open pods, and is well presented for combine harvesting. Seed yield is usually very good, despite 50 per cent of the plants producing no seed. However, crop duration is often limited by senescence and eventual infestations by larvae of the Lychnis Moth (*Hadena bicruris*). Separation of Red Campion seed from Lychnis Moth larval faeces is a challenge at cleaning.

194. *Silene flos-cuculi*: Ragged Robin (native)

Features: Ragged Robin (former Latin name: *Lychnis flos-cuculi*) is a bisexual perennial of wet places. It is widespread and frequent in wet meadows, rush pasture, ditches, dune slack and tall herb fen.

Seed production: Ragged Robin establishes readily in cultivation, doing better on wetter ground, and is not dioecious. The fine seed is held loose at maturity in

Ragged Robin (*Silene flos-cuculi*).

Seed of Ragged Robin.

upward-pointing, small open capsules, and may be collected by hand or combine harvested.

195. *Silene latifolia*: White Campion (archaeophyte)

Features: White Campion (former Latin name: *Silene alba*) is a wintergreen perennial/biennial and is frequent in hedgerows, waysides and wasteland, and sometimes on cultivated ground. White Campion is dioecious, with half of all plants being female (with large flowers), and half male (with smaller flowers). It is night scented, moth pollinated, and susceptible to Lychnis Moth infestations.

Seed production: In cultivation, the greyish seed germinates best from an autumn drilling, benefitting from vernalisation, and seed crops establish well on good ground. Only the female plants produce seed, and this seed is retained loose in near upright, open capsules. Mature plants are readily combine harvested, producing good yields of seed, comparable to that of Red Campion. Red and White Campion readily hybridise when grown in close proximity, producing intermediate pink-flowered hybrids (*Silene* x *hampaena*).

White Campion (*Silene latifolia*).

Seed of White Campion.

196. *Silene uniflora*: Sea Campion (native)

Features: Sea Campion (former Latin name: *Silene maritima*) is a cushion-forming perennial halophyte of mainly coastal locations on shingle, drift lines, cliffs and maritime grasslands.

Sea Campion (*Silene uniflora*) growing on Chesil Bank, Dorset, with Kidney Vetch and Thrift.

Seed production: It may be grown in cultivation, but will require careful attention to weed control, and most often the seed will be harvested by collecting ripe pods by hand.

197. *Silene vulgaris:* Bladder Campion (native)

Features: Bladder Campion is a long-lived deciduous perennial of waysides, hedge banks, rough pasture and other grassy places on a wide range of soils. It has simple grey/green leaves and white flowers with a swollen and veined calyx tube.

Seed production: The seed has no dormancy restrictions, germinating readily from an early autumn or spring drilling to produce neat rows. Plants die back below ground in winter, building stores in an extensive fleshy root system.

RIGHT: Bladder Campion (*Silene vulgaris*).

BELOW: Seed of Bladder Campion.

Winter weed growth can be checked by shallow spring-tine cultivation. Green shoots sprout above ground in mid-spring, and growth up to flowering is rapid. Crops are usually healthy but sometimes susceptible to stem rot. Bladder Campion is not dioecious, so all plants form seeds, and at maturity these seeds are retained loose in hard-walled pods within the swollen and persistent calyx tube. Harvesting by combine harvester requires a closed drum to shatter the seedpods, and moderate aspiration. Bladder Campion is not susceptible to the Lychnis Moth and seed yields can be very good.

198. *Spiranthes spiralis*: Autumn Lady's-tresses (native)

Features: Autumn Lady's-tresses is a widespread and infrequent, rosette-forming, tuberous perennial Orchid of short vegetation in grassy places. Flowering is in late summer, and the inflorescence consists of an unbranched one-sided spike of small white flowers, usually twisted in a spiral.

Autumn Lady's-tresses (*Spiranthes spiralis*).

Seed production: Seed of Autumn Lady's-tresses should not be collected from the wild, but may be collected, with permission if needed, when it is found growing, as it occasionally does, in lawns. It may sometimes be used for restoration within its native range.

199. *Stachys sylvatica*: Hedge Woundwort (native)

Features: Hedge Woundwort is a widespread and common rhizomatous perennial of hedges, waysides, woodland, wasteland and riverbanks. For anyone who has grown up playing in the countryside as a child, memories will be stirred by the unique, musty aroma of crushed Hedge Woundwort leaves.

RIGHT: Hedge Woundwort (*Stachys sylvatica*).

BELOW: Seed of Hedge Woundwort.

Seed production: Hedge Woundwort is an important restoration species because it has a place in a wide range of habitats and is certainly attractive. In cultivation it can form clonal patches, but seed harvest can be tricky and yield low. It does not combine harvest well because seed is dropped from the plant soon after maturity. It is better to gather maturing inflorescences and dry them on a sheet.

200. *Stellaria holostea*: Greater Stitchwort (native)

Features: Greater Stitchwort is a common, wintergreen perennial of hedges, clearings in deciduous woodland, and waysides. It has narrow leaves tapering to a point, and the five petals of the white flowers are notched to half way.

Seed production: Greater Stitchwort may be propagated using horticultural methods. The plants grow well, and pollinated flowers produce spherical seedpods containing seeds that turn orange when mature. The seedpods may be gathered by hand and dried to release the seed.

201. *Succisa pratensis*: Devil's-bit Scabious (native)

Features: Devil's-bit Scabious is an important restoration species for meadows and is widespread in many other habitats. It is a variable perennial with moderately large, entire basal leaves and ascending inflorescences, having many individual flowers consisting of clusters of blue florets. It is now much less common than formerly. Devil's-bit Scabious is so variable in form that ecotypes confined to isolated meadows may be easily distinguished one from another.

Seed production: Seed of Devil's-bit Scabious can be precision drilled in rows, and this seed will readily germinate and grow strongly to produce mature seed, producing plants in the second full year after drilling. Short rows may be hand harvested by plucking seed heads every few days as they mature. Larger, field-scale crops can be combine harvested at peak seed maturity.

Greater Stitchwort (*Stellaria holostea*).

Devil's-bit Scabious (*Succisa pratensis*).

Seed of Greater Stitchwort.

Seed of Devil's-bit Scabious.

202. *Taraxacum officinale*: Dandelion (native)

Features: Dandelion is a tap-rooted, wintergreen, rosette-forming, short-lived, ubiquitous perennial. It is completely apomictic, meaning that it produces seed by

Dandelion (*Taraxacum officinale*).

Dandelion clock.

Dandelion seed.

somatic division of cells without pollination. The attractive flowers do not need to attract pollinators, nor do they need to provide food rewards in the form of pollen and nectar, but they do. Dandelion is one of the top five native species for the provision of pollinator services in the form of pollen and nectar (Hicks et al., 2016). Dandelion seed is buff coloured, ribbed and toothed at the distal end, just below the point of attachment of a slender beak, at the far end of which is attached a white feathered pappus.

Seed production: Dandelion is a fairly determinate species, producing a major flush of flowers in late April and a major flush of seed maturation in early May. This early seeding, combined with wind dispersal of seed, allows the species to colonise early cut grassland destined for silage harvesting. These grasslands can provide an opportunity for Dandelion seed harvesting. The Dandelion seed harvesting period is confined to two weeks in early May. Wind in this period can remove the Dandelion seed from the field before it is harvested, and on average this happens in three years out of every four. After harvest, the seed must be rubbed to break the beak to which the pappus is attached, and this should be done before the seed is dried and cleaned.

203. *Teucrium scorodonia*: Wood Sage (native)

Features: Wood Sage is a rhizomatous perennial of well-drained acidic to mildly calcareous soils across a wide range of habitats in both the open and semi-shade. The crushed leaves have a hemp-like scent, and the yellow/green flowers produce seeds that are retained within the calyx.

Wood Sage (*Teucrium scorodonia*).

Seed production: Wood Sage can be grown in cultivation, and seed harvest is by gathering mature inflorescences. After drying, the seed can be released from the material by rubbing hard.

204. *Thalictrum flavum*: Common Meadow-rue (native)

Features: Common Meadow-rue is a rhizomatous perennial of damp meadows, fens, ditches and streamsides. Once common, it is now far less frequent.

Seed production: Common Meadow-rue is slow to establish from seed and best suited to small-scale horticultural propagation. Seed may be harvested at maturity by hand collection.

Common Meadow-rue (*Thalictrum flavum*).

205. *Thymus drucei*: Wild Thyme (native)

Features: Wild Thyme is an evergreen, dwarf, long-lived perennial with creeping stems. It is occasional on free-draining calcareous soils in rocky places, on anthills, heath, downland and sea cliffs, sand dunes and in upland grasslands up to and over 1,000m (3,300ft).

Wild Thyme (*Thymus drucei*).

Seed production: Wild Thyme may be grown from seed on suitable substrates using horticultural methods. Establishment of plants is slow, and seed may be harvested by plucking mature inflorescences. On drying, the fine seed can be separated, but normally only small quantities will be obtained.

206. *Torilis japonica*: Upright Hedge-parsley (native)

Features: Upright Hedge-parsley is a winter annual, or biennial umbellifer of neutral to calcareous soils. It is summer flowering, and is occasionally common in hedgerows, waysides and rough grassland.

Seed production: Upright Hedge-parsley may be precision drilled in rows in late summer/early autumn, and will crop in the following year if the seed germinates before the winter. Harvest is by combine harvester, producing good yields of seed.

Upright Hedge-parsley (*Torilis japonica*).

207. *Tragopogon pratensis*: Goat's Beard (native)

Features: Goat's Beard (Jack-go-to-bed-at-noon) is a short-lived perennial of meadows, pasture, waysides and unmanaged grasslands. The grass-like leaves usually have a pale mid-rib, and the yellow capitula close at midday. The seeds are rather like those of an over-sized Dandelion, being large, buff coloured, slightly curved, and having a long slender beak and a huge terminal feathered pappus.

Seed production: The seed of Goat's Beard is best collected by hand from open seeding inflorescences. The

Goat's Beard (*Tragopogon pratensis*).

species may be grown in cultivation, or seed may be collected from wayside or meadow populations, where it is permitted to do so.

208. *Trifolium campestre*: Hop Trefoil (native)

Features: Hop Trefoil is a winter annual of dry, infertile calcareous grasslands. It is common and widespread, but rather less so than Lesser Trefoil. It is distinguished by having flowers with the standard deflexed.

Seed production: Seed of Hop Trefoil is sometimes obtained as a component of brush-harvested seed and because of its small size can be sorted from most other species.

Hop Trefoil (*Trifolium campestre*).

209. *Trifolium dubium*: Lesser Trefoil (native)

Features: Lesser Trefoil is a very common winter annual of meadows, pastures, waysides, lawns and amenity grassland. In flower it resembles Black Medick (*Medicago lupulina*), but in seed it is distinguished from Black Medick by not having clusters of black seeds. It is further distinguished from Hop Trefoil in having the standard petal folded over the flower. It also has a much wider habitat range tolerance.

Seed production: Like Hop Trefoil, seed of Lesser Trefoil may sometimes be obtained from brush-harvested mixtures, especially those from a mid-June harvest.

Lesser Trefoil (*Trifolium dubium*).

210. *Trifolium pratense*: Red Clover (native)

Features: Wild native Red Clover is an ascending perennial, abundant in meadows, waysides and rough grasslands. It is a very variable species, showing diversity of form in habit, stipules, leaf, flower colour and other features. Wild Red Clover provides pollinator services and is an important restoration species of meadows. It is grazing intolerant and is largely replaced by White Clover (*Trifolium repens*) in pastures. Agricultural cultivars of Red Clover are extensively grown as a nitrogen-fixing forage legume, and they tend to be more robust and shorter lived compared to the wild type.

Seed production: Wild Red Clover may be precision drilled in rows at a field scale. The crop establishes readily and will often display huge variation in features within the crop. The seed is retained well on the inflorescences at maturity, and may be harvested by combine harvester with the concave closed and aspiration moderate. Red Clover is a proscribed species and is subject to the Seed Marketing Regulations 2011. Regardless, wild-type native ecotypes are preferred for restoration seeding.

Red Clover (*Trifolium pratense*).

211. *Trifolium repens*: White Clover (native)

Features: Wild native White Clover is an abundant and ubiquitous stoloniferous creeping perennial of pasture, lawns, amenity grassland and most other short grasslands. It is very tolerant of grazing, mowing and trampling. White Clover is often present in meadows, but during growth of the hay crop it usually persists in a semi-dormant state at the base of the sward. During this period Red Clover is normally dominant over White Clover. Mowing in July removes the above-ground biomass that has been shading the White Clover. This often results in explosive growth of White Clover and an out-of-season burst of flowering in late summer, while the aftermath is being grazed.

Seed production: Seed of wild-type native White Clover may be produced using the same methods as are used to produce agricultural White Clover seed. White Clover is a proscribed species subject to the Seed Marketing Regulations 2011; it is an important nitrogen-fixing forage legume. Agricultural cultivars are widely sown, obscuring the distribution of the native wild type. Agricultural cultivars tend to be more robust and shorter lived than the native wild type. White Clover is an important restoration species, and it is better to use the native wild type when restoring and creating habitat.

White Clover (*Trifolium repens*).

212. *Trisetum flavescens*: Yellow Oat-grass (native)

Features: Yellow Oat-grass is a tufted perennial of well-drained neutral and calcareous grasslands. It is frequent and common in meadows, pastures, downland and on waysides in central and southern regions, but less so in the far north. The inflorescences are fine and feathery, and golden in colour.

Yellow Oat-grass (*Trisetum flavescens*).

Seed of Yellow Oat-grass.

Seed production: Yellow Oat-grass may be precision drilled in rows at a field scale, producing seed crops that are suitable for direct combine harvesting with aspiration set low.

213. *Typha latifolia*: Bulrush (native)

Features: Bulrush is a familiar rhizomatous perennial of the water edge beside ponds, canals and ditches, and also, but less frequently, beside moving water. The inflorescence is terminal on a tall, unbranched stem of 2–3m (6.5–10ft), and consists of a cigar-like tight cluster of female flowers below a loose tassel of male flowers. The female flowers produce seed that is fine and numerous. At maturity the seed spike splits and shatters, liberating the seed that is then wind dispersed, aided by hair-like seed appendages.

Seed production: Mature inflorescence spikes of Bulrush are readily collected by hand from waterside locations in late summer. The seed is released from the main stem on drying and may be separated from non-seed material by hand sorting. The hair-like appendages are best left attached to the seed.

Bulrush (*Typha latifolia*).

Seed of Bulrush.

214. *Umbilicus rupestris*: Navelwort (native)

Features: Navelwort is a short-lived perennial of rocky places on the coast and inland. The plant has a basal rosette of succulent rounded leaves with a near-central shallow depression marking the point of attachment to the petiole. The leaf depression has the appearance of a naval. The inflorescence stem is unbranched and has a long terminal spike of greenish flowers. These mature to produce dust-like brownish/black seeds.

Navelwort (*Umbilicus rupestris*).

Seed production: Seeding inflorescences may be collected by hand from plants in cultivation, and drying releases the dust-like seed. Seed yield is often very small.

215. *Verbascum nigrum*: Dark Mullein (native)

Features: Dark Mullein is a biennial or short-lived perennial of well-drained calcareous soils. It is locally frequent on disturbed soils and wasteland in central and southern Britain. The strikingly yellow flowers with violet anther filaments are borne in loose spikes on branched inflorescences.

Seed production: Cropping plants may be raised in rows by precision drilling in the autumn or spring. Flowering takes place in the first or second full year after seeding, and maximum seed yield is obtained by cutting seeding inflorescences as seed capsules mature, and drying on sheets to release the seed.

Dark Mullein (*Verbascum nigrum*).

216. *Verbascum thapsus*: Great Mullein (native)

Features: Great Mullein is a biennial of well-drained neutral to calcareous soils. It is widespread and occasional on disturbed ground, waysides, waste ground, quarries and railways in central and southern Britain. Great Mullein produces a rosette of large, powdery-looking leaves and a tall, usually unbranched, inflorescence spike with tight-packed yellow flowers.

Seed production: Great Mullein may be precision drilled in rows, and the species is cultivated and harvested in a way similar to that for Dark Mullein. Great Mullein is clothed in irritant down and it is wise to take extra precautions when handling the plant and cleaning the seed. Seed yield can be very good.

Large soft rosette leaves of Great Mullein (*Verbascum thapsus*).

217. *Veronica chamaedrys*: Germander Speedwell (native)

Features: Germander Speedwell is a stoloniferous, clump-forming perennial frequent on anthills, waysides, hedgerows, woodland rides and downland. It is common and attractive, spreading vegetatively to form patches.

Germander Speedwell (*Veronica chamaedrys*).

Seed production: Small-scale plots of Germander Speedwell may be established from seed or rooted stolons. Seed capsules may be harvested in bunches by hand and dried to release the seed. Seed yield is usually very poor.

218. *Veronica persica*: Field Speedwell (naturalised neophyte)

Features: Field Speedwell was first recorded as an introduction in 1825 in Berkshire. It is a self-fertile annual of cultivated and disturbed ground and is now common, abundant and frequent throughout the British Isles. Field Speedwell produces large quantities of seed and forms a persistent seed bank in the soil.

Seed production: Field Speedwell may be precision drilled in the autumn, or it may spring naturally as a first-year weed in drillings of a perennial crop. It grows too low to pick up with a combine harvester, but may be raked into swathes with spring tine harrows. On drying, the swathes can be picked up by a combine harvester with aspiration and concave both set at medium. Seed yield can be very good.

Field Speedwell (*Veronica persica*).

219. *Vicia cracca*: Tufted Vetch (native)

Features: Tufted Vetch is a handsome scrambling or climbing perennial Vetch of meadows, pasture, waysides, hedgerows and scrub. The leaves are greyish/green, once pinnate, and ladder-like with narrow leaflets and a terminal tendril. The vivid blue flowers are held in upright clusters, and after pollination the flattened pods hang, turning from green to light brown. The seed is hard coated and requires scarification to break dormancy.

Tufted Vetch (*Vicia cracca*).

Seed of Tufted Vetch.

Seed production: Field-scale row crops may be established by precision drilling in the autumn or spring. Establishment is rather slow, with first flowering in the second or third full year after drilling. The crop may be grown with the support of accompanying upright weeds such as Creeping Thistle (*Cirsium arvense*), or equally without support. Mature pods are retained on the crop for several weeks, and may be harvested directly by combine harvester with the concave closed and aspiration up. Seed yield is good, and Tufted Vetch will continue to crop for several years.

220. *Vicia sativa*: Common Vetch (native)

Features: Common Vetch is a pink-flowered climbing winter annual of waysides, hedgerows and grassy places. The leaves are once pinnate with a terminal tendril. Summer leaves are large and have elongated, slightly rounded leaflets; winter rosette leaves are small and have quite rounded and rather crinkly leaflets. The flowers are

Common Vetch (*Vicia sativa*).

Seed of Common Vetch.

in pairs (rarely singly) on short pedicels in leaf axils, and on pollination produce a pair of long, round-section pods pointing upwards in the shape of a 'V'. At first green, the pods turn dark brown, then black, just before explosively splitting to discharge seeds.

Seed production: Seed of Wild Common Vetch may be broadcast or precision drilled in the early autumn or spring to produce field-scale crops. Flowering is in the first full year after drilling, and harvest, which is timing critical, is by combine harvester with the conclave closed and aspiration high. Seed yield is very good. Common Vetch is a proscribed species subject to the Seed Marketing Regulations 2011, and agricultural cultivars are frequently sown for green manure, herbal forage and game cover, often escaping into the wild.

221. *Vicia sepium*: Bush Vetch (native)

Features: Bush Vetch is a spring-flowering, climbing perennial of hedges, waysides, scrub, woodland edge and rough, grassy places. The leaves are once pinnate with rounded leaflets and a terminal tendril.

Seed production: Bush Vetch may be grown in cultivation, and a combine harvester may harvest the seed if the plants are cultivated in sufficient quantity. Smaller plots may be harvested by collecting mature pods and drying these to release the seed.

Bush Vetch (*Vicia sepium*).

222. *Viola arvensis*: Field Pansy (archaeophyte)

Features: Field Pansy is an opportunistic spring annual with a short life cycle of three to four months, allowing populations to survive on disturbed and cultivated ground that is not treated with herbicides.

Seed production: Field Pansy may be grown in cultivation, and will often arise naturally from the soil seed bank. The plants are too low-growing for seed to be harvested by combine, but mature plants may be gathered and dried on sheets. The seed can be released from the dry pods by crushing.

Field Pansy (*Viola arvensis*).

Common Dog-violet (*Viola riviniana*).

223. *Viola riviniana*: Common Dog-violet (native)

Features: Common Dog-violet is a woodland perennial, also frequent on waysides, hedge banks, meadows, heath, moorland, cliffs and rocky places. The leaves are heart shaped, and the largish flowers are blue with a pale spur. Plants are much branched at the base and form clumps.

Seed production: Common Dog-violet can be grown in cultivation from seed at a horticultural scale. Flowers are borne individually on pedicels, and after pollination the pedicels corkscrew downwards to place the seed capsules beneath the leaves to assist seed dispersal by foraging invertebrates. The capsules may be gathered by hand at maturity from beneath the leaves, before they are robbed of seed.

224. *Viola tricolor*: Wild Pansy (native)

Features: Wild Pansy is an increasingly rare annual/biennial of sandy soils on heathland, sand dunes, field margins and disturbed ground. The leaves are elongate with divided stipules, and the flowers, larger than those of Field Pansy, are attractive, multicoloured and variable in pigmentation.

Seed production: Wild Pansy may be precision drilled in rows at a field scale and will usually flower in the first full year after drilling. Flowering and seed production can be prolific, but the seedpods are not sufficiently well presented for combine harvesting. They may, however, be gathered in bunches and dried on a sheet to release the seed.

APPENDIX

Table 2: Sowing Rates (g/m^2) for Restoration Species in Habitat Mixtures

RESTORATION SPECIES	Status (1)	Seeds/g (2)	Annual (3)	Neutral (4)	Calcar-eous (5)	Sandy (6)	Wetland (7)	Upland (8)	Moor Heath (9)	Wayside Tussock (10)	Hedge Scrub (11)	Wood (12)	Coast (13)	Production 2023 (kg) (14)
Achillea millefolium	N	6,000	-	0.02	0.01	0.03	-	-	-	0.01	trace	-	0.03	300
Achillea ptarmica	N	4,000	-	-	-	-	trace	-	-	-	-	-	-	10
Adonis annua	A	500	0.02	-	-	-	-	-	-	-	-	-	-	1
Aethusa cynapium	N	250	0.06	-	---	-	-	-	-	-	-	-	-	10
Agrimonia eupatoria	N	50	-	-	-	-	-	-	-	0.14	0.14	-	0.10	150
Agrostemma githago	A	70	0.40	-	-	-	-	-	-	-	-	-	-	3,000
Agrostis capillaris	N	15,000	-	0.01	-	0.02	0.01	0.02	0.02	-	-	-	0.04	xxx
Ajuga reptans	N	700	-	-	-	-	0.01	-	-	-	0.02	0.01	-	1
Alliaria petiolata	N	500	-	-	-	-	-	-	-	-	0.06	0.02	-	200
Allium ursinum	N	150	-	-	-	-	-	-	-	-	-	0.20	-	100
Alopecurus geniculatus	N	1,000	-	-	-	-	0.02	-	-	-	-	-	-	50
Alopecurus pratensis	N	1,000	-	0.04	-	-	0.06	-	-	0.04	0.01	-	-	500
Anacamptis morio	N	250,000	-	trace	trace	-	trace	-	-	-	-	-	-	---
Anacamptis pyramidalis	N	250,000	-	-	trace	trace	-	-	-	-	-	-	trace	---
Anemone nemorosa	N	450	-	-	-	-	-	-	-	-	-	0.01	-	---
Angelica sylvestris	N	160	-	-	-	-	0.10	-	-	-	0.04	-	-	50
Anthoxanthum odoratum	N	1,500	-	0.06	0.04	0.04	0.08	0.04	0.04	-	-	-	0.10	300
Anthriscus sylvestris	N	300	-	-	-	-	-	-	-	0.10	0.10	0.05	-	100
Anthyllis vulneraria	N	400	-	-	0.01	0.06	-	-	-	-	-	-	0.08	300
Armeria maritima	N	700	-	-	-	-	-	-	-	-	-	-	0.02	10
Arrhenatherum elatius	N	300	-	-	-	-	-	-	-	0.01	-	-	-	xxx
Arum maculatum	N	20	-	-	-	-	-	-	-	-	0.10	0.20	-	10
Asplenium scolopendrium	N	---	-	-	-	-	-	-	-	-	trace	trace	-	---
Avenella flexuosa	N	3,000	-	-	-	0.05	-	0.10	0.20	-	-	-	-	200
Avenula pubescens	N	320	-	-	0.02	-	-	-	-	-	-	-	-	100
Barbarea vulgaris	N	1,300	0.01	-	-	-	-	-	-	-	-	-	-	100
Bellis perennis	N	10,000	-	trace	-	-	-	-	-	-	-	-	-	1
Betonica officinalis	N	700	-	0.05	-	-	0.08	-	-	-	-	-	-	250

RESTORATION SPECIES	Status (1)	Seeds/g (2)	Annual (3)	Neutral (4)	Calcar-eous (5)	Sandy (6)	Wetland (7)	Upland (8)	Moor Heath (9)	Wayside Tussock (10)	Hedge Scrub (11)	Wood (12)	Coast (13)	Production 2023 (kg) (14)
Blackstonia perfoliata	N	20,000	-	-	trace	trace	-	-	-	-	-	-	trace	1
Brachypodium sylvaticum	N	400	-	-	-	-	-	-	-	0.08	1.00	0.06	-	100
Briza media	N	2,500	-	0.10	0.08	-	0.06	-	-	-	-	-	-	400
Bromopsis erecta	N	200	-	-	0.25	-	-	-	-	0.10	0.04	-	-	500
Calluna vulgaris	N	10,000	-	-	-	-	-	trace	0.04	-	-	-	-	20
Caltha palustris	N	1,000	-	-	-	-	trace	-	-	-	-	-	-	20
Campanula glomerata	N	8.000	-	-	trace	-	-	-	-	-	-	-	-	20
Campanula rotundifolia	N	15,000	-	trace	trace	trace	-	trace	trace	-	-	-	trace	20
Campanula trachelium	N	4,200	-	-	-	-	-	-	-	-	0.01	0.01	-	100
Capsella bursa-pastoris	A	6,000	trace	-	-	-	-	-	-	-	-	-	-	10
Cardamine pratensis	N	2,000	-	-	-	-	trace	-	-	-	trace	trace	-	---
Carduus nutans	N	500	-	-	0.02	-	-	-	-	-	-	-	-	20
Carex divulsa	N	800	-	-	-	-	-	-	-	0.08	0.10	-	-	150
Carex echinata	N	600	-	-	-	-	0.04	-	0.10	0.08	-	-	-	150
Carex flacca	N	800	-	0.04	0.04	0.02	0.02	0.04	-	-	-	-	-	100
Carex pendula	N	1,300	-	-	-	-	0.04	-	-	-	-	-	-	30
Carex sylvatica	N	800	-	-	-	-	-	-	-	-	-	0.30	-	20
Carlina vulgaris	N	900	-	-	trace	trace	-	-	-	-	-	-	trace	1
Centaurea cyanus	A	250	0.40	-	-	-	-	-	-	-	-	-	-	1,500
Centaurea nigra	N	450	-	0.08	0.04	0.06	0.08	0.08	-	0.08	0.06	-	0.06	1,400
Centaurea scabiosa	N	160	-	-	0.10	-	-	-	-	0.04	-	-	0.04	400
Centaurium erythraea	N	80,000	-	-	trace	trace	-	-	-	-	-	-	trace	1
Cerastium fontanum	N	3,000	trace	trace	trace	-	-	-	-	-	-	-	-	20
Chaenorhinum minus	A	10,000	trace	-	-	-	-	-	-	-	-	-	-	2
Chaerophyllum temullum	N	500	-	-	-	-	-	-	-	0.10	0.18	0.14	-	100
Circaea lutetiana	N	600	-	-	-	-	-	-	-	-	0.02	0.01	-	2
Cirsium acaule	N	400	-	-	0.01	-	-	-	-	-	-	-	-	2
Cirsium eriophorum	N	200	-	-	0.03	-	-	-	-	-	-	-	-	25
Cirsium palustre	N	1,000	-	-	-	-	trace	-	trace	-	-	-	-	10
Clematis vitalba	N	300	-	-	-	-	-	-	-	-	0.02	-	-	30
Clinopodium vulgare	N	2,400	-	-	-	-	-	-	-	-	0.02	-	-	60
Conopodium majus	N	1,200	-	0.02	0.02	-	-	-	-	-	-	0.02	-	20
Crambe maritima	N	15	-	-	-	-	-	-	-	-	-	-	0.20	10
Crithmum maritimum	N	250	-	-	-	-	-	-	-	-	-	-	0.10	10

RESTORATION SPECIES	Status (1)	Seeds/g (2)	Annual (3)	Neutral (4)	Calcar-eous (5)	Sandy (6)	Wetland (7)	Upland (8)	Moor Heath (9)	Wayside Tussock (10)	Hedge Scrub (11)	Wood (12)	Coast (13)	Production 2023 (kg) (14)
Cruciata laevipes	N	200	-	-	-	-	-	-	-	0.18	0.24	-	0.24	250
Cynoglossum officinale	N	20	-	-	-	0.20	-	-	-	-	-	-	0.10	25
Cynosurus cristatus	N	1,000	-	2.00	1.80	1.60	2.00	1.60	1.00	0.20	0.20	-	2.00	1,000
Dactylis glomerata	N	800	-	0.04	-	-	0.04	-	-	0.15	0.15	-	-	xxx
Dactylorhiza fuchsii	N	250,000	-	trace	trace	trace	trace	trace	-	trace	trace	-	trace	---
Dactylorhiza praetermissa	N	250,000	-	trace	-	-	trace	trace	trace	-	-	-	trace	---
Daucus carota	N	1,000	-	0.06	-	0.10	-	-	-	0.10	0.02	-	0.04	xxx
Deschampsia cespitosa	N	4,000	-	-	-	-	0.05	-	-	0.10	0.10	0.20	-	500
Digitalis purpurea	N	13,000	-	-	-	-	-	-	0.02	-	-	0.06	0.02	300
Dipsacus fullonum	N	300	-	-	-	-	-	-	-	0.12	0.06	-	-	500
Dryopteris felix-mas	N	---	-	-	-	-	-	-	trace	-	-	trace	-	---
Echium vulgare	N	250	-	-	-	0.08	-	-	-	-	-	-	-	300
Epipactis heleborine	N	250,000	-	-	-	-	-	-	-	-	-	trace	-	---
Epipactis palustris	N	250,000	-	-	-	-	trace	-	-	-	-	-	-	---
Erica cinerea	N	50,000	-	-	-	-	-	trace	0.04	-	-	-	-	20
Erica tretrahit	N	50,000	-	-	-	-	-	-	0.04	-	-	-	-	20
Ervilia sylvatica	N	100	-	-	-	-	-	-	-	-	-	0.14	-	---
Eryngium maritimum	N	50	-	-	-	-	-	-	-	-	-	-	0.15	30
Eupatorium cannibinum	N	4,000	-	-	-	-	trace	-	-	-	-	-	-	20
Euphorbia exigua	A	400	trace	-	-	-	-	-	-	-	-	-	-	---
Euphorbia helioscopia	A	300	trace	-	-	-	-	-	-	-	-	-	-	---
Festuca ovina	N	1,200	-	-	0.25	0.40	-	0.50	0.60	-	-	-	0.50	xxx
Festuca rubra	N	1,000	-	0.25	0.10	0.30	0.40	0.30	0.40	0.25	0.20	-	0.30	xxx
Filipendula ulmaria	N	1,000	-	-	-	-	0.04	-	-	0.04	0.04	-	-	400
Filipendula vulgaris	N	1,000	-	0.01	0.08	-	0.01	-	-	-	-	-	-	150
Fragaria vesca	N	1,500	-	-	-	-	-	-	-	-	trace	trace	-	1
Fritillaria meleagris	NN	250	-	-	-	-	0.05	-	-	-	-	-	-	40
Fumaria officinalis	A	300	0.05	-	-	-	-	-	-	-	-	-	-	30
Galeopsis tetrahit	N	250	0.01	-	-	-	-	-	-	-	-	-	-	10
Galium album	N	1,200	-	-	0.04	-	0.02	-	-	0.08	0.12	-	-	500
Galium odoratum	N	1,200	-	-	-	-	-	-	-	-	-	0.04	-	---
Galium verum	N	3,200	-	0.04	0.08	0.10	-	0.08	0.10	-	-	-	0.10	1,200
Genista tinctoria	N	300	-	0.04	-	-	0.04	-	-	-	-	-	-	10
Geranium dissectum	A	800	0.02	0.02	0.01	0.01	-	-	-	-	-	-	-	40

RESTORATION SPECIES	Status (1)	Seeds/g (2)	Annual (3)	Neutral (4)	Calcar-eous (5)	Sandy (6)	Wetland (7)	Upland (8)	Moor Heath (9)	Wayside Tussock (10)	Hedge Scrub (11)	Wood (12)	Coast (13)	Production 2023 (kg) (14)
Geranium lucidum	N	1,000	0.01	-	-	0.01	-	-	-	-	0.01	-	-	5
Geranium molle	N	1,000	0.01	-	-	0.01	-	-	-	-	-	-	-	5
Geranium pratense	N	100	-	-	-	-	-	-	-	0.10	0.10	-	-	300
Geranium pyreniacum	NN	600	-	-	-	-	-	-	-	-	0.06	-	-	100
Geranium robertianun	N	750	-	-	-	-	-	-	-	-	0.01	0.01	-	10
Geum rivale	N	700	-	-	-	-	0.02	0.01	-	-	-	-	-	30
Geum urbanum	N	400	-	-	-	-	-	-	-	0.04	0.04	0,04	-	100
Glaucium flavum	N	4,000	-	-	-	-	-	-	-	-	-	-	0.02	2
Glebionis segetum	A	600	0.15	-	-	-	-	-	-	-	-	-	-	800
Helianthemum nummularium	N	900	-	-	0.01	0.01	-	-	-	-	-	-	0.01	10
Heracleum sphondylium	N	150	-	0.10	0.05	-	-	-	-	0.10	0.08	-	-	50
Hippocrepis comosa	N	350	-	-	0.02	0.02	-	-	-	-	-	-	0.02	100
Hordeum secalinum	N	200	-	0.15	0.10	-	-	-	-	-	-	-	-	300
Hyacinthoides non-scripta	N	120	-	-	-	-	-	0.10	-	-	0.10	0.40	0.10	300
Hypochaeris radicata	N	1,500	-	0.02	0.02	0.02	-	0.02	0.02	-	-	-	0.02	50
Iris foetidissima	N	20	-	-	-	-	-	-	-	-	-	0.40	-	30
Iris pseudacorus	N	20	-	-	-	-	0.20	-	-	-	-	-	-	100
Jasione montana	N	50,000	-	-	-	-	-	-	-	-	-	-	trace	1
Kickxia elantine	A	---	trace	-	-	-	-	-	-	-	-	-	-	1
Kickxia spuria	A	---	trace	-	-	-	-	-	-	-	-	-	-	1
Knautia arvensis	N	150	-	0.10	-	0.10	-	-	-	0.10	0.06	-	-	500
Koeleria macrantha	N	3,000	-	-	0.05	0.05	-	0.05	-	-	-	-	-	200
Lamiastrum galeobdolon	N	350	-	-	-	-	-	-	-	-	0.01	0.01	-	5
Lamium album	A	1,000	-	-	-	-	-	-	-	-	0.02	-	-	5
Lamium purpureum	A	1,200	0.02	-	-	-	-	-	-	-	-	-	-	10
Lapsana communis	N	1,000	-	-	-	-	-	-	-	-	0.01	-	-	10
Lathyrus pratensis	N	60	-	0.12	0.10	-	0.12	-	-	0.10	0.08	-	-	200
Lathyrus sylvestris	N	25	-	-	-	-	-	-	-	-	0.10	-	0.20	20
Legousia hybrida	A	10,000	trace	-	-	-	-	-	-	-	-	-	-	1
Leontodon hispidus	N	900	-	0.04	0.04	0.04	-	0.04	0.04	-	-	-	0.04	400
Leucanthemum vulgare	N	2,000	-	0.08	0.08	0.01	0.08	0.04	-	0.08	0.06	-	0.08	1,000
Linaria vulgaris	N	4,000	-	-	-	-	-	-	-	0.01	-	-	-	5
Linum bienne	N	800	0.05	0.02	0.04	-	-	-	-	-	-	-	-	100

RESTORATION SPECIES	Status (1)	Seeds/g (2)	Annual (3)	Neutral (4)	Calcar-eous (5)	Sandy (6)	Wetland (7)	Upland (8)	Moor Heath (9)	Wayside Tussock (10)	Hedge Scrub (11)	Wood (12)	Coast (13)	Production 2023 (kg) (14)
Linum catharticum	N	10,000	-	trace	trace	trace	-	-	-	-	-	-	-	5
Lotus corniculatus	N	800	-	0.06	0.08	0.10	0.04	0.04	0.08	-	-	-	0.08	xxx
Lotus pedunculatus	N	800	-	-	-	-	0.01	-	-	-	-	-	-	100
Lycopus europeus	N	4,000	-	-	-	-	trace	-	-	-	-	-	-	10
Lysimachia arvensis	N	3,500	0.03	-	-	-	-	-	-	-	-	-	-	40
Lythrum salicaria	N	20,000	-	-	-	-	trace	-	-	-	-	-	-	50
Malva moschata	N	500	-	-	0.08	0.10	-	-	-	0.10	-	-	-	500
Malva sylvestris	A	160	-	-	-	-	-	-	-	0.04	0.04	-	-	200
Matricaria chamomilla	A	10,000	0.02	-	-	-	-	-	-	-	-	-	-	100
Matricaria discoidea	NN	10,000	0.01	-	-	-	-	-	-	-	-	-	-	20
Medicago lupulina	N	400	-	0.08	0.08	0.10	-	-	-	-	-	-	-	xxx
Mentha aquatica	N	5,000	-	-	-	-	trace	-	-	-	-	-	-	5
Mercurialis perennis	N	500	-	-	-	-	-	-	-	-	0.02	0.04	-	5
Myosotis arvensis	A	2,000	0.03	-	-	-	-	-	-	-	-	-	-	50
Neottia ovata	N	250,000	-	-	-	-	-	-	-	-	trace	trace	-	---
Odontites vernus	N	5,200	0.01	-	-	-	-	-	-	-	-	-	-	40
Oenanthe pimpinelloides	N	450	-	0.04	0.06	-	-	-	-	0.08	-	-	-	100
Onobrychis viciifolia	(N)	25	-	-	0.10	-	-	-	-	-	-	-	-	xxx
Ononis repens	N	160	-	-	0.06	-	-	-	-	-	-	-	-	20
Ophioglossum vulgatum	N	---	-	trace	trace	-	trace	-	-	-	trace	-	-	---
Ophrys apifera	N	250,000	-	trace	trace	trace	-	-	-	-	trace	-	trace	---
Orchis mascula	N	250,000	-	-	-	-	-	-	-	-	-	trace	trace	---
Origanum vulgare	N	11,000	-	-	0.02	-	-	-	-	-	0.01	-	-	100
Orobanche minor	N	100,000	-	trace	trace	trace	-	-	-	-	-	-	-	1
Papaver dubium	A	6,000	0.05	-	-	-	-	-	-	-	-	-	-	50
Papaver rhoeas	A	10,000	0.30	-	-	-	-	-	-	-	-	-	-	2,000
Paris quadrifolia	N	---	-	-	-	-	-	-	-	-	-	0.02	-	---
Pastinaca sativa	N	230	-	-	0.05	-	-	-	-	0.12	-	-	-	50
Phleum arenarium	N	4,000	-	-	-	-	-	-	-	-	-	-	0.02	4
Phleum bertolonii	N	4,000	-	0.10	0.10	0.10	-	0.10	-	-	-	-	-	xxx
Pilosella officinarum	N	6,000	-	-	trace	trace	-	trace	-	-	-	-	trace	1
Pimpinella major	N	500	-	-	-	-	-	-	-	0.08	0.06	-	-	50
Pimpinella saxifraga	N	900	-	0.04	0.06	0.02	-	-	-	-	-	-	0.02	100
Plantago coronopus	N	3,500	-	-	-	-	-	-	-	-	-	-	0.02	1

RESTORATION SPECIES	Status (1)	Seeds/g (2)	Annual (3)	Neutral (4)	Calcar-eous (5)	Sandy (6)	Wetland (7)	Upland (8)	Moor Heath (9)	Wayside Tussock (10)	Hedge Scrub (11)	Wood (12)	Coast (13)	Production 2023 (kg) (14)
Plantago lanceolata	N	300	-	0.10	0.04	0.08	0.10	0.08	-	-	0.02	-	0.08	1,200
Plantago media	N	2,000	-	-	0.04	-	-	-	-	-	-	-	-	400
Platanthera chlorantha	N	250,000	-	-	trace	-	-	-	-	-	-	-	-	---
Polygala vulgaris	N	---	-	-	trace	-	-	-	-	-	-	-	-	---
Polypodium vulgare	N	---	-	-	-	-	-	-	-	-	-	trace	-	---
Poterium sanguisorba	N	250	-	-	0.25	0.06	-	-	-	-	-	-	-	600
Primula veris	N	900	-	0.06	0.06	0.04	0.05	0.05	-	0.04	0.05	-	0.05	800
Primula vulgaris	N	1,300	-	-	-	-	-	-	-	-	-	0.01	-	10
Prunella vulgaris	N	1,200	-	0.08	0.08	0.06	0.06	0.06	-	-	0.06	-	0.06	1,200
Pulicaria dysenterica	N	5,000	-	-	-	-	trace	-	-	-	-	-	-	5
Ranunculus acris	N	600	-	0.12	0.10	-	0.14	0.14	-	0.10	0.10	-	-	1,400
Ranunculus arvensis	A	180	0.10	-	-	-	-	-	-	-	-	-	-	60
Ranunculus auricormus	N	600	-	-	-	-	-	-	-	-	-	0.01	-	5
Ranunculus bulbosus	N	350	-	0.04	0.08	-	-	0.04	-	-	-	-	-	200
Reseda lutea	N	1,350	-	-	0.04	-	-	-	-	-	-	-	-	100
Rhinanthus minor	N	300	-	0.50	0.50	0.50	0.50	0.50	0.50	0.50	0.50	-	0.50	1,500
Rubia perregrina	N	150	-	-	-	-	-	trace	trace	-	-	-	trace	1
Rumex acetosa	N	1,400		0.08	0.04	-	0.08	-	-	0.02	0.01	-	-	400
Rumex acetosella	N	1,500	-	-	-	0.04	-	0.02	0.06	-	-	-	-	100
Salvia verbenaca	N	700	-	-	-	-	-	-	-	0.04	-	-	-	20
Sanguisorba officinalis	N	600	-	-	-	-	0.06	0.01	-	0.01	-	-	-	200
Sanicula europeus	N	300	-	-	-	-	-	-	-	-	-	0.02	-	2
Saxifraga granulata	N	35,000	-	trace	-	-	-	-	-	-	-	-	-	1
Scabiosa columbaria	N	800	-	-	0.05	-	-	-	-	-	-	-	-	200
Scrophularia auriculata	N	12,000	-	-	-	-	0.01	-	-	-	-	-	-	5
Sedum acre	N	25,000	-	-	-	trace	-	-	-	-	-	-	trace	1
Serratula tinctoria	N	1,100	-	0.04	-	-	0.01	-	-	-	-	-	-	2
Sherardia arvensis	N	2,000	0.02	-	-	-	-	-	-	-	-	-	-	5
Silaum silaus	N	400	-	-	-	-	0.08	-	-	-	-	-	-	100
Silene dioica	N	1,000	-	-	-	-	-	-	-	-	0.12	0.25	-	700
Silene flos-cuculi	N	5,000	-	-	-	-	0.01	-	-	-	-	-	-	300
Silene latifolia	A	800	-	-	-	-	-	-	-	0.08	-	-	-	500
Silene uniflora	N	1,500	-	-	-	-	-	-	-	-	-	-	0.06	20
Silene vulgaris	N	1,200	-	-	0.06	0.04	-	-	-	0.08	0.10	-	-	400

RESTORATION SPECIES	Status (1)	Seeds/g (2)	Annual (3)	Neutral (4)	Calcar-eous (5)	Sandy (6)	Wetland (7)	Upland (8)	Moor Heath (9)	Wayside Tussock (10)	Hedge Scrub (11)	Wood (12)	Coast (13)	Production 2023 (kg) (14)
Spiranthes spiralis	N	---	-	-	trace	-	-	-	-	-	-	-	-	---
Stachys sylvatica	N	600	-	-	-	-	-	-	-	-	0.02	0.02	-	20
Stellaria holostea	N	450	-	-	-	-	-	-	-	-	0.02	0.02	-	5
Succisa pratensis	N	650	-	-	-	-	0.06	-	-	-	-	-	-	300
Taraxacum officinale	N	1,300	-	0.05	0.05	0.05	0.05	0.05	-	-	-	-	0.05	200
Teucrium scorodonia	N	1,400	-	-	-	0.04	-	0.01	0.04	-	-	0.04	0.04	20
Thalictrum flavum	N	600	-	-	-	-	0.01	-	-	-	-	-	-	20
Thymus drucei	N	8,000	-	-	trace	trace	-	trace	-	-	-	-	trace	10
Torillis japonica	N	600	-	-	-	-	-	-	-	0.10	0.08	-	-	100
Tragopogon pratensis	N	60	-	0.10	0.10	0.02	-	-	-	0.02	0.02	-	-	60
Trifolium campestre	N	2,000	-	trace	trace	trace	-	-	-	-	-	-	-	5
Trifolium dubium	N	2,000	-	trace	trace	trace	-	-	-	-	-	-	-	10
Trifolium pratense	N	500	-	0.08	0.06	-	0.08	0.08	-	-	0.04	-	-	xxx
Trifolium repens	N	1,400	-	0.02	0.02	0.02	-	0.02	-	-	-	-	-	xxx
Trisetum flavescens	N	2,500	-	0.10	0.10	-	-	-	-	-	-	-	-	200
Typha latifolia	N	---	-	-	-	-	trace	-	-	-	-	-	-	5
Umbilicus rupestris	N	50,000	-	-	-	-	-	-	-	-	-	-	trace	1
Verbascum nigrum	N	10,000	-	-	-	-	-	-	-	0.01	-	-	-	30
Verbascum thapsus	N	12,000	-	-	-	0.01	-	-	-	-	-	-	0.01	100
Veronica chamaedris	N	3,800	-	-	-	-	-	-	-	trace	trace	-	-	2
Veronica persica	N	3,000	0.04	-	-	-	-	-	-	-	-	-	-	100
Vicia cracca	N	40	-	0.20	0.20	-	0.20	-	-	0.15	0.15	-	-	400
Vicia sativa	N	50	-	-	-	-	-	-	-	0.15	0.15	-	-	xxx
Vicia sepium	N	35	-	-	-	-	-	-	-	-	0.20	0.10	-	10
Viola arvensis	A	2,000	0.04	-	-	-	-	-	-	-	-	-	-	10
Viola riviniana	N	1,000	-	-	-	-	-	-	-	-	0.02	0.02	-	5
Viola tricolor	N	1,400	0.04	-	-	-	-	-	-	-	-	-	0.04	50

Key to columns in Table 2

(1) Status: N=native, A=archaeophyte, NN=naturalised neophyte.

(2) Seeds/g: Estimates derived from counts and published data (Henniges et al., 2021).

(3-13) Sowing rates: Standard rates (g/m^2) are provided for species when included in different habitat mixtures. Species inclusion and sowing rate will vary depending on the time of year, project objectives, companion species, site conditions, availability. The habitat columns correspond with those habitats described in the text. For each habitat, sowing rates for core species are coloured blue, and those for optional additional species are coloured green.

(14) Production 2023 (kg): Estimate of the total production of individual species in Britain and Ireland, excluding brush harvested mixtures. No data is given for Proscribed Species (xxx).

GLOSSARY

achene A one-seeded fruit.
actinomorphic A flower with radial symmetry.
adventive Introduced but not fully naturalised.
alien Introduced deliberately or accidentally by man; not native.
allomone A type of semiochemical produced and released by an individual of one species that affects the behaviour of another species.
allopolyploid Having more than two sets of chromosomes derived from different ancestral species.
ambophily Regular distribution of pollen by both insects and wind.
androecium Male parts of the flower.
anenophily Regular distribution of pollen by wind.
annual Life cycle of twelve months or less.
anther The pollen-bearing part of a stamen.
anthesis Pollen shedding.
apomictic The production of seed without fertilisation.
appendage A small extension.
archaeophyt Naturalised species introduced by humans intentionally or unintentionally between the start of the neolithic period and AD1489.
auricle An ear-shaped appendage.
autogamous Self-pollination of a flower.
awn A stiff bristle.
beak A narrow apical projection.
biennial A life cycle of more than one but less than two years.
biodeposition Deposition of bio-organic material.
bisexual A plant or flower bearing both sexes.
bract A scale-like leaf.
brush harvester A machine that harvests seeds by brushing.
calcareous Substrate containing calcium carbonate.
calcicole A species confined to lime-rich soil.
calcifuge A species intolerant of lime-rich soil.
calyx The sepals of a flower.
capitulum A dense, flat cluster of small flowers or florets.
carpel The female reproductive organ of flowering plants consisting of ovary (style if present) and stigma.
casual An alien plant that escapes cultivation but doesn't persist in the wild.
clonal Non-sexual reproduction.
community A cluster of individuals of different species.
corolla Petals and enclosed reproductive organs.
corymb A cluster of flowers forming a flattened inflorescence.
cotyledon The first leaves of a plant pre-formed within the seed.
cross-pollination The pollination between individuals.
cultivar A variety produced by selective artificial breeding.
cyme A branched inflorescence with the central terminal bud opening first.
deciduous Leaves are shed in the autumn and leafless in the winter.
deflexed Bent or curving downwards or backwards.
dehiscence The opening of a seedpod.
denizen A naturalised archaeophyte.
determinate A short, concentrated flowering period.
dimorphic Occurring in two forms.
dioecious Having the male and female reproductive organs on separate individuals.
diploid Having two complete matching sets of chromosomes, 2n.
disc Anything disc shaped.
dormancy The state of having normal functions suspended.
drupe A fleshy fruit.
ecocline A gradual and continuous environmental gradation.
ecotype A genetically distinct population within a species.
edaphic Influenced by the soil.
elaiosome Fleshy structures externally attached to seeds serving to attract vectors.
endemic Restricted to one particular area.
endosperm The part of the seed that acts as a food store for the developing embryo.
entomophily Pollination by insects.
epicalyx Organs on the outside of the flower that are calyx-like and in addition to the calyx.
epiphyte A plant growing on another plant.
extinction vortex An accelerating extinction in small populations.
filament Stalk of a stamen.
fecundity Fruitfulness and fertility.
floret A small flower.
gametophyte The haploid generation of a plant.
gene flow The movement of genetic information between populations.
genome The entire set of DNA instructions found in a cell.
genomics A discipline focusing on the structure, evolution and editing of genomes.
genotype The genetic constitution of an organism.
genus A taxonomic rank above species but below family.
germination The process by which an organism grows from a seed or spore.
glaucus Covered with a greyish, bluish or whitish waxy coating or bloom.
guttation The secretion of drops of zylem sap from the leaf tips.
gynodioecious Having female and bisexual/hermaphrodite plants within the same species.
gynoecium The female reproductive parts of a flower.
halophyte A salt-tolerant species.
haploid Having only one set of chromosomes.
haustoria A root-like structure that grows into or around another plant.
haylage Grass cut earlier than hay and wilted to 5–40 per cent moisture content.
hectare 10,000 square metres.
hemi-parasitic A green, partially parasitic species.
herbaceous Non-woody ground flora.
heterophyllus Plants having leaves of two or more different forms.

heterosis Hybrid vigour.
heterostylus A species with two or three morphological types of flower.
heterozygous Having two different alleles of a particular gene.
hilum The scar on a seed left by its point of attachment.
hybrid A cross between plants of two different taxa.
hyphae The branching filaments of a fungus.
imbibed Fully hydrated.
inbreeding Breeding genetically related individuals.
indehiscent Seedpod remaining intact and not opening.
indeterminate Continuing to flower indefinitely.
individual A single organism.
inflorescence The complete flower head of a plant.
introduced A plant that has been imported into a country by the human hand.
introgression The transfer of genetic information from one species to another as a result of hybridisation between them and repeated backcrossing.
invasive A species with the ability to colonise new territory.
involucre A whorl of bracts below an inflorescence.
labellum A modified petal, usually larger than the other petals.
laciniate A leaf divided deeply.
ligulate Strap-shaped.
local From a nearby location.
locus A specific position on a chromosome.
longevity Duration of survival.
meiosis Reduction division of chromosomes to produce gametes.
mesotrophic Of soils with a moderate inherent fertility.
micro-climate A fine scale local set of climatic conditions.
micro-species A distinct genotype that is reproduced by apomixes.
mitosis Somatic nuclear duplication and division.
monocarpic Plants that flower and set seed only once and then die.
morphotype A group of different and distinct individuals in a population.
mutant A genetic variant.
mutualism The beneficial interaction between two or more species.
mycorrhiza Soil fungi intimately associated with plant roots.
myxospermous Seed having a covering of sticky mucilage.
native Having colonised a place by natural means.
naturalised Reproducing and self-sustaining in the wild.
nectariferous Producing nectar.
neophyte An alien species.
non-native A species whose natural native range is elsewhere.
nutlets A small, nut-like seed.
obligate An organism that is restricted to a certain way of life.
octoploid Having eight sets of chromosomes.
origin The place in the wild from which stock has come.
orthodox Seeds that can be stored dry for long periods.
outbreeding Crossing between genetically different individuals.
ovary The basal part of the gynoecium containing the ovules.
ovule The part of the ovary that contains the female germ cell.
pappus A modified persistent calyx made up of a ring of fine hairs, scales or teeth.
parasite An organism that draws nourishment from a host.
pedicel Stalk of a flower.
peduncle Stalk of a group of flowers.
perennate To live over from one growing season to another.
perennial A plant that lives more than two years.
perianth The outer, non-sexual parts of the flower.
petaloid Resembling the petal of a flower.
petals Modified leaves that surround the reproductive parts of the flower.
petiole The stalk of a leaf.
phenotype The observable characteristics of an organism.
phytogeographic The geographic distribution of plants.
pinnae The smallest division of a divided leaf.
pinnate A compound leaf having leaflets on opposite sides of the midrib.
plastic Varying in form according to environmental conditions.
ploidy The number of chromosome sets in a cell.
pollen The male gametes of a flower.
pollinia A coherent mass of pollen grains.
polycarpic Plants that flower and set seed many times before dying.
polyploidy Where the cells of an organism have more than one set of chromosomes.
population A freely interbreeding group of individuals.
provenance The place of cultivation and production of reproductive material.
pubescent Soft down or fine short hairs on the leaves or stems of plants.
radicle The embryonic root of a plant.
ray Strap-like marginal flowers.
recalcitrant Seeds that do not survive drying.
receptacle The thickened part of the stem from which the flower organs grow.
rhizome A modified underground or ground-level stem.
saprophyte A plant deriving its nutrition from decaying organisms.
scarification Weakening of the seed coat.
seed A fertilised ovule.
self-compatible Self-fertile.
self-incompatible Self-sterile.
senesce Biological ageing.
sepal One of the segments of the outer whorl(s) of the perianth.
sessile Not stalked.
silage Fermented grass with a moisture content of between 40 and 60 per cent.
somatic Any cell that forms the body of a multicellular organism other than a gamete.
sorus A cluster of sporangia on a fern frond.
spadix Minute flowers on a fleshy stem.
spathe A leaf-like curved bract that typically ensheaths the spadix.
species A basic unit of biological classification.
speciation The process of generating species.
spermatophyte Any plant that produces seed.
spores Haploid reproductive cells.
sporophyte The diploid generation of a plant.
spring annual Annual that normally germinates in the spring.
spur A protrusion of any part of a flower.

stamen The pollen-producing male part of a flower.
stigma The apical tip of the gynoecium that is receptive to pollen.
stipule One of (usually) a pair of appendages at the base of a leaf or its stalk.
stolon A creeping horizontal stem that takes root.
style The stalk of a stigma.
sub-species A rank below species.
sward An area covered with grass.
symbiosis A close mutualistic biological interaction between two species.
taxon Any taxonomic grouping.
tepal Petal-like sepals.
terracette A step-like landform seen on steep ground created by livestock.
testa The outer coat of a seed.
tetraploid Having four sets of chromosomes, 4n.
tiller A shoot that rises from the base of a grass plant.
transpiration Evaporation from leaves pulling water through a plant.
triploid Of a plant which has three sets of chromosomes, 3n.
umbel A plate-like inflorescence.
unisexual Of a plant or flower bearing only one sex.
venation The arrangement of veins in a leaf.
vernalisation Subjection of fully imbibed seed to cold temperatures to break dormancy.
viability Alive.
viscidium The sticky part of the rostellum in an orchid.
weed A species that is judged to cause harm.
winter annual An annual species that is winter hardy and will germinate in the autumn.
wintergreen A plant that overwinters above ground and remains green.
zygomorphic Of a flower with bilateral symmetry.
zygote A fertilised egg cell formed by the fusion of two gametes.

REFERENCES

Abrahamczyk S., Struck J-H., Weigend M. (2023). Pollination mode and reproductive system of *Sanguisorba minor* and *Sanguisorba officinalis*. *Plant Species Biology* 38:4, 171–179.

Adamatzky A., (2022). Language of fungi derived from their electrical spiking activity. *Royal Society Open Science* 9:4, 1–15.

Albrecht H., Cambecedes J., Lang M., Wagner M. (2016). Management options for the conservation of rare arable plants in Europe. *Botany Letters* 163:4, 389–415.

Anten N.P.R. and Chen B.J.W. (2021). Detect thy family: Mechanisms, ecology and agricultural aspects of kin recognition in plants. *Plant Cell & Environment* 44:4, 1059–1071.

Arroyo-Correa B., Burkle L.A., Emer C. (2019). Alien plants and flower visitors disrupt the seasonal dynamics of mutualistic networks. *Journal of Ecology.* 108:4, 1475–1486.

Auffret G.A., Rico Y., Bullock J.M., Hooftman A.P., Pakeman R.J., Soons M.B., Suarez-Esteban A., Traveset A., Wagner H.H., Cousins S.A.O. (2017). Plant functional connectivity – integrating landscape structure and effective dispersal. *Journal of Ecology* 105:6, 1648–1656.

Baldock Pypass (2009). Monitoring Report 2009. *Hertfordshire Biological Records Centre.*

Bardgett R.D., Marsden J.H. and Howard D.C. (1995). The extent and condition of heather moorland in the uplands of England and Wales. *Biological Conservation*, 71:2, 155–161.

Barke B.H., Daubert M., Horandl E. (2018). Establishment of apomixis in dipoloid F2 hybrids and inheritance of apospory from F1 and F2 hybrids of the *Ranunculus auricomus* complex. *Frontiers in Plant Science* 9: 1–12.

Barraclough T.G. and Humphreys A.M. (2015). The evolutionary reality of species and higher taxa in plants: a survey of post-modern opinion and evidence. *New Phytologist* 207: 291–296.

Berthon K., Thomas F., Bekessy S. (2021). The role of 'nativeness' in urban greening to support animal biodiversity. *Landscape and Urban Planning* 205: 1–11.

Bialic-Murphy L., Brewer N.L., Kaliz S. (2019). Direct effects of a non-native invader erode native plant fitness in the forest understory. *Journal of Ecology* 108:1, 189–198.

Bilas R.D., Bretman A., Bennett T. (2020). Friends, neighbours and enemies: an overview of the communual and social biology of plants. *Plant Cell & the Environment* 44:4, 997–1013.

Bonanomi G., Zotti M., Idbella M., Termolino P., De Micco V., Mazzoleni S. (2022). Field evidence for litter and self-DNA inhibitory effects on *Alnus glutinosa* roots. *New Phytologist* 236:2, 399–412.

Bowers J.K. (1985). In: British Agricultural Policy since the Second World War.

Breed M.F., Stead M.G., Ottewell K.M., Gardner M.G., Lowe A.J. (2013). Which provenance and where? Seed-sourcing strategies for revegetation in a changing environment. *Conservation Genetics* 14: 1–10.

Brooker R.W., George T.S., Homulle Z., Karley A.J., Newton A.C., Pakeman R.J., Schob C. (2021). Facilitation and biodiversity–ecosystem function relationships in crop production systems and their role in sustainable farming. *Journal of Ecology* 109:5, 2054–2067.

Broughton R.K., Bullock J.M., George C., Gerard F., Maziarz M., Payne W.E., Scholefield P.A., Wade D., Pywell R.F. (2022). Slow development of woodland vegetation and bird communities during 33 years of passive rewilding in open farmland. *PLoS ONE* 17:(11).

Brown M.R., Moore P.G.P., Twyford A.D. (2021). Performance of generalist hemiparasitic Euphrasia across a phylogenetically diverse host spectrum. *New Phytologist* 232:5, 2165–2174.

Brys L.R. and Jacquemyn H. (2009). Biological Flora of the British Isles: *Primula veris*. *Journal of Ecology*, 97: 581–600

Brys R., Jacquemyn H., Endels P., De Blust G. and Hermy M. (2004). The effects of grassland management on plant performance and demography in the perennial herb Primula veris. *Journal of Applied Ecology* 41: 1080–1091.

Bucharova A., Michalski S., Hermann J-M, Heveling K., Durka W., Holzel N., Kollmann J., Bossdorf O. (2017). Genetic differentiation and regional adaptation among seed origins used for grassland restoration: lessons from a multispecies transplant experiment. *Journal of Applied Ecology* 54:1, 127–136.

Bucharova A., Bossdorf O., Hölzel N., Kollmann J. (2019). Mix and match: regional admixture provenancing strikes a balance among different seed-sourcing strategies for ecological restoration. *Conservation Genetics* 20:1, 7–17.

Bullock J.M., Pywell R.F. and Walker K.J. (2007). Long-term enhancement of agricultural production by restoration of biodiversity. *Journal of Applied Ecology* 4:1, 6–12.

Butler L.H., Hay F.R., Ellis R.H., Smith R.D., Murray T.B. (2009). Priming and re-drying improve the survival of mature seeds of *Digitalis purpurea* during storage. *Annals of Botany* 103:8, 1261–1270.

CABI digital library. (2019). On *Rhododendron ponticum* as an invasive species. (2019). CABI compendium. Datasheet 581–600, CABI Head Office, Wallingford, UK.

Carver C., Convery I., Hawkins S., Beyers R., Eagle A., Kun Z., Maanen E.V., Cao Y., Fisher M., Edwards S.R., Nelson C., Gann G.D., Shurter S., Aguilar K., Andrade A., Ripple W.J., Davis J., Sinclair A., Bekoff M., Noss R., Foreman D., Pettersson H., Root-Bernstein M., Svenning J-C., Taylor P., Wynne-Jones S., Featherstone A.W., Flojgaard C., Stanley-Price M., Navarro L. M., Aykroyd T., Parfitt A., Soule M. (2021). Guiding principles for rewilding. *Conservation Biology.* 35:6, 1882–1893.

Clapham A.R., Tutin T.G., Warburg E.E. (1962). *Flora of the British Isles.*

Cong W-F., Ruijven J., Mommer L., De Deyen G.B., Berdense F., Hoffland E. (2014). Plant species richness promotes soil carbon and nitrogen stocks in grasslands without legumes. *Journal of Ecology* 102: 1163–1170.

Conrady M., Lampei C., Bossdorf O., Holzel N., Michalski S., Durka W., Bucharva A. (2023). Plants cultivated for ecosystem restoration can evolve towards a domestication syndrome. *PNAS* 2023 120: 20 1–7.

Corbet S.A., Bee J., Dasmahapatra K., Gale S., Gorringe E., La Ferla B., Moorhouse T., Trevail A., Van Bergen Y., Vorontsova (2001). Native or Exotic? Double or Single? Evaluating Plants for Pollinator-friendly Gardens. *Annals of Botany* 87: 219–223.

Cosyns E., Hoffmann M. (2005). Horse-dung germinable seed content in relation to plant species abundance, diet composition and seed characteristics. *Basic and Applied Ecology* 6: 11–24.

Coverdale T., Davies A.B. (2023). Unravelling the relationship between plant diversity and vegetational structural complexity. A review and theoretical framework. *Journal of Ecology* 111:7, 1378–1395.

Cowley J.M., Burton M.A. (2020). The good stuff: Plantago as a myxospermous model with modern utility. *New Phytologist* 229:4, 1917–1923.

Darwin C. (1859). *On the origin of species.*

Darwin C. (1876). *The effects of gross and self-fertilization in the vegetable kingdom.*

Davies A. (2002). The Floristic Restoration of Agriculturally Improved Grassland. University of Sheffield, *Thesis submitted for the Degree of Doctor of Philosophy,* 2002.

Davies K., Leger E.A. (2021.) Living with exotic annual gasses in the Sagebush ecosystem. *Journal of Environmental Management* 288.

Defra. (2004). *Code of Practice on how to prevent the Spread of Ragwort* 2004.

Egan P.A., Stevenson P.C., Tiedeken E.J., Wright G., Boylan F., Stout J. (2006). Plant toxin levels in nectar vary spatially across native and introduced populations. *Journal of Ecology* 104: 1106–1115.

Ercole E., Adamo M., Rodda M., Gebauer G., Girlanda M., Perotto S. (2014). Temporal variation in mycorrhizal diversity and carbon and nitrogen stable isotope abundance in the wintergreen meadow orchid *Anacamptis morio. New Phytologist* 205:3, 1308–1319.

Erickson V.J., Halford A. (2020). Seed planning, sourcing, and procurement. *Restoration Ecology* 28:S3 219–227.

Evert F.K., Cockburn M., Beniers J.E., Latsch R. (2020). Weekly defoliation controls, but does not kill Broad-leaved Dock (*Rumex obtusifolius*) *Weed Research.* DOI: 10.1111/wre.12407.

Farrell C.A., Aronson J., Daily G.C., Hein L., Obst C., Woodworth P., Stout J.C. (2022). Natural capital approaches: shifting the UN Decade on Ecosystem Restoration from aspiration to reality. *Restoration Ecology* 30:7, 1–6.

Farrell L. (1993). Lowland Heathland: The extent of habitat change. *English Nature Science* No.12. English Nature Peterborough.

Feniuk C., Balmford A., Green R.E. (2019). Land sparing to make space for species dependent on natural habitats and high nature value farmland. *Proc Biol Sci B* 286, 1–7.

Fenner M., Thompson K. (2005). *The Ecology of Seeds.* Cambridge University Press.

Filipiak. M. (2019). Key pollen host plants provide balanced diets for wild bee larvae: A lesson for planting flower strips and hedgerows. *Journal of Applied Ecology* 56:6, 1410–1418.

Floodplain Meadows Partnership News Letter (2013). Issue No. 10 January winter 2012/13.

Flory S.L., Bauer J., Phillips R.P., Clay K. (2017). Effects of a non-native grass invasion decline over time. *Journal of Ecology* 105:6, 1475–1484.

Frischie S., Miller A.L., Pedrini S., Kildisheva O.A. (2020). Ensuring seed quality in ecological restoration: native seed cleaning and testing. *Restoration Ecology* 28:S3, 239–248.

Fuller R.M. (1987). The changing extent and conservation interest of lowland grasslands in England and Wales: A Review of Grassland Surveys 1930–1984. *Biological Conservation Volume* 40:4, 281–300.

Gamble D., Perry C., St Pierre T. (2012). Hay Time, Final Report 2012. Yorkshire Dales.

Garibaldi L.A., Oddi F.J., Miguez F.E., Bartomeus I., Orr M.C., Estoban G.J., Kremen K., Schulte L.A., Hughes A.C., Bagnato C., Abramson G., Bridgewater P., Gomez Carella D., Diaz S., Dicks L.V., Ellis E.C., Goldenberg M., Huaylla C.A., Kuperman M., Locke H., Mehrabi Z., Santibanez F., Zhu C-D. (2020). Working Landscapes need at least 20% native habitat. *Conservation Letters* 14:2.

Geert A., Rossum F. and Ludwig Triest L. (2010). Do linear landscape elements in farmland act as biological corridors for pollen dispersal? *Journal of Ecology* 98:1, 178–187.

Gewin V. (2020). The zoologist tracking an island's rebirth. How horses and cows are enticing skylarks and starlings back to an island meadow. *Nature* 588(7838):530.

Gopsill M.A., and Farooque G.A. (2023). Incorporated cover crop residue suppresses weed seed germination. *Weed Biol. Manag.* 23:2, 1–10.

Gosney R.F.C., Gellatley G., Kerr A., Petch K.A. and Peck R.B. (1999). A4/A46 Batheaston/Swainswick Bypass. *Proceedings of the Institute of Civil Engineering.* 120:3, 101–110.

Guillemin J.P., Bellanger S., Reibeil C., Darmency H. (2017). Longevity, dormancy and germination of *Cyanus segetum. Weed Research* 57:6, 361–371.

Hancock N., Gibson-Roy P., Driver M. and Broadhurst L. (2020). The Australian Native Seed Sector Survey Report. Australian Network for Plant Conservation, Canberra.

Hanggi E., Ingersoll J.F. (2012). Lateral vision in horses: A behavioral investigation. *Behavioural Processes* 91:1, 70–76.

Havens K., Vitt P., Kramer A.T., Fant J.B., Schatz K. (2015). Seed sourcing in an era of climate change. *Natural Areas Journal* 35, 122–133.

Hayhow D.B., Eaton M., Burns F., Stanbury A. (2019). *The State of Nature 2019.* The State of Nature Partnership.

Heckman R.W., McColley C., Slater M.N., and Carr D.E. (2016). The role of community composition in grassland response to two methods of exotic forb removal. *Weed Research* 57:1, 44–53.

Henniges M.C., Powell R.F., Mian S., Stace C.A., Walker K.J., Gornall R.J., Christenhusz M.J.M., Brown M.R., Twyford A.D., Hollingsworth P.M., Jones L., De Vere N., Antonelli A., Leitch A.R., Leitch I.J. (2021). *A taxonomic, genetic and ecological data resource for the vascular plants of Britain and Ireland.* NERC Environmental Information Data Centre.

Hicks D.M., Ouvrard P., Baldock K.C.R., Baude M., Goddard M.A., Kunin W.E. (2016). Food for Pollinators: Quantifying the Nectar and Pollen Resources of Urban Flower Meadows. *PLoS ONE* 11(6): e0158117.

Hill S.M., Vila-Aiub M., Hernandez M., Kaspary T.E., Garcia M.A. (2022). Cross- and multiple herbicide resistance in *Lolium multiflorum* across Uruguay. *Weed Research* 62:4, 296–305.

Hofmann M.M., Renner S.S. (2020). One-year-old flower strips already support a quarter of a city's bee species. *Journal of Hymenoptera Research* 75: 87–95.

Houston K., Wolff K. (2012). *Rhinanthus minor* population genetic structure and subspecies: Potential seed sources of a keystone species in grassland restoration projects. *Perspectives in Plant Ecology, Evolution and Systematics* 14: 423–433.

Hume C. (2008). Wetland Vision Technical Document: overview and reporting of project philosophy and technical approach. *The Wetland Vision Partnership.*

Irwin A. (2023). Can the Netherlands' artificial islands lead the way to rewilding the planet? *Nature* 616: (7958), 644–648.

Isbell F., Adler P.R., Eisenhauer N., Fornara D., Kimmel K., Kremen C., Letourneau D.K., Liebman M., Polley H.W., Quijas S., Scherer-Lorenzen M. (2017). Benefits of increasing plant diversity in sustainable agroecosystems. *Journal of Ecology,* 105:4, 871–879.

Isermann M. and Rooney P. (2014). Biological Flora of the British Isles: *Eryngium maritimum. Journal of Ecology,* 102, 789–821

Jacquemyn H., Brys R., Merckx V.S.F.T., Waud M., Lievens B., Wiegand T. (2013). Coexisting Orchid species have distinct mycorrhizal communities and display strong spatial segregation. *New Phytologist* 202:2, 616–627.

Jacquemyn H., Brys R. and Hutchings M.J. (2014). Biological Flora of the British Isles: *Epipactis palustris. Journal of Ecology* 102: 1341–1355.

Jefferson R.G. and Walker K.J. (2017). Biological Flora of the British Isles: *Serratula tictoria. Journal of Ecology* 105: 1438–1458.

Jefferson R.G. et al. (2023). Biological Flora of the British Isles: *Geranium pratense. Journal of Ecology* 111:12, 2750–2780.

JNCC (2022). *Nature Recovery for Our Survival, Prosperity and Wellbeing.* A joint statement by the Statutory Nature Conservation Bodies of the UK.

Johnson D. and Gilbert L. (2015). Interplant signalling through hyphal networks. *New Phytologist* 205:4, 1448–1453.

Johnson L.C., Galliart M.B., Alsdurf J.D., Maricle B.R., Baer S.G., Bello N.M., Gibson D.J., Smith A.B. (2021). Reciprocal transplants gardens as a gold standard to detect local adaption in grassland species: New opportunities moving into the 21st century. *Journal of Ecology* 110:5, 1054–1071.

Johnson W.E., Onorato D.P., Roelke M.E., Darrel Land E., Cunningham M., Belden R. C., Mcbride R.,Jansen D., Lotz M., Shindle D., Howard J., Wildt D.E., Penfold L.M., Hosteler J.A., Oli M.K., O'Brien S.J. (2010). Genetic Restoration of the Florida Panther. *Science* 329:59, 1641–1645.

Jones T.A. (2013). When local isn't best. *Evolutionary Applications.* 6:7, 1109–1118.

Joshi J. *et al* (2001). Local adaptation enhances performance of common plant species. *Ecology Letters* 4: 536–544.

Keble Martin. (1965). *Concise British Flora in Colour.* ISBN:9780718140281.

Kedzierski G. (2013). US native seed growing. Native seed production for grassland restoration - a grower's perspective. *Ecological Restoration* 32:2, 130–133.

Kiehl K., Kirmer A., Tischew S., Shaw N., editors. (2014). *Guidelines for Native Seed Production and Grassland Restoration.* Cambridge Scholars Publishing. ISBN (13):979–14438–5900–4.

Kildisheva O., Dixon K., Silveira F.A.O. (2020). Dormancy and germination: Making every seed count. *Restoration Ecology* 28:53, 256–265.

Knapp E., Rice K.J. (1994). Starting from Seed. Genetic Issues in Using Native Grasses for Restoration. *JSTOR Restoration and Management Notes* 12:1, 40–45.

Kohn D.D., Ruhsam M., Hulme P.E., Barrett S.,C.H., Hollingsworth P.M. (2019). Paternity analysis reveals constraints on hybridization potential between native and introduced bluebells (*Hyacinthoides*). *Conservation Genetics* p1-14 https//doi.org/10.1007/s10592-019-01158-4

Kotilinek M., Tesitlova T., Jersakova J. (2015). Biological Flora of the British Isles: *Neottia ovata. Journal of Ecology.*

Ladouceur E., Isbell F., Clark A.T., Harpole W.S., Reich P.B., Tilman G.D., Chase J.M. (2023). The recovery of plant community composition following passive restoration across spatial scales. *Journal of Ecology* 111:4, 814-829.

Landis D.A. (2017). Designing agricultural landscapes for biodiversity-based ecosystem services. *Science Direct* 18: 1–12.

Lang M., Prestele J., Fischer C., Kollman J., Albrecht H. (2016). Reintroduction of rare arable plants by seed transfer. What are the optimal sowing rates? *Ecology and Evolution* 6:15, 5506-5516.

Lange M., Roth V-N., Eisenhaur N., Roscher C., Dittmar N., Fischer–Bedtke C., Gonzalez Mace O., Hildebrandt A., Milcu A., Mommer L., Oram N.J., Ravenek J., Scheu S., Schmid B., Strecker T., Wagg C., Weigelt A., Gleixner G. (2020). Plant diversity enhances production of downward transport of biodegradable dissolved organic matter. *Journal of Ecology* 109:3, 1284–1297.

Lauterbach D., Burkart M., Gemeinholzer B. (2012). Rapid genetic differentiation between *ex situ* and their *in situ* source populations: an example of the endangered *Silene otitis* (Caryophyllaceae). *Botanical Journal of the Linnean Society* 168: 64–67.

Lawson C., Rothero E., Gowing D., Nisbet T., Barsoum N., Broadmeadow S., Skinner A. (2018). The natural capitol of floodplains: management, protection and restoration to deliver greater benefits. *Valuing Nature: Natural Capital Synthesis Report.*

Lawton J.H., Brotherton P.N.M., Brown V.K., Elphick C., Fitter A.H., Forshaw J., Haddow R.W., Hilborne S., Leafe R.N., Mace G.M., Southgate M.P., Sutherland W.J., Tew T.E., Varley J., Wynne G.R. (2010). *Making Space for Nature: a review of England's wildlife sites and ecological network.* Report to Defra.

Lenton T.M., Crouch M., Johnson M., Pires N., Dolan L. (2012). First plants cooled the Ordovician. *Nature Geoscience* 5: 86–89.

Li J., Luo X., Zhou K. (2024). Research and development of hybrid rice in China. *Plant Breeding* 143: 96–104.

Li Y., Stift M., Kleunen M. (2017). Admixture increases performance of an invasive plant beyond first-generation heterosis. *Journal of Ecology* 106:4, 1596–1606.

Lobo K., Mardon A.A., Johnson J.C., Johnson P.A. (2021). On Genetic Rescue: A Brief Conspectus. *Academia Letters* 2021.

Lopasio G., Hasday E.N., Espadlier X., Germann C., Ortiz-Sanchez F.J., Pont A., Sommaggio D., Schob C. (2021). Facilitation and biodiversity jointly drive mutualistic networks. *Journal of Ecology* 109:1.

MacIntyre D. and Elliott C.G. (1974). Selection for growth rate during asexual and sexual propagation in *Phytophthora cactorum. Genetic Research, Cambridge.* 24: 295–309.

MacIntyre D. (2006). Plant Rings. *BSBI News* 101: 10–11.

MacIntyre D. (2014). An interesting cluster of Orchid species. *BSBI News* 126: 9–10.

Maclaren C., Bennett J., Dehen-Schmutz K. (2019). Management practices influence the competitive potential of weed communities and their value to biodiversity in South African vineyards. *Weed Research* 59:2, 93–106.

Mainz A.K. and Wieden M. (2019). Ten years of native seed certification in Germany – a summary. *Plant Biology* 21: 383–388.

Marrs R.H., Watt A.S. (2006). Biological Flora of the British Isles: *Pteridium aquilinum* (L.) Kuhn. *Journal of Ecology* 94: 6, 1272–1321.

Marshall and Brown in Frankel O.H., Hawkes J.G. (1975). *Crop Genetic Resources for Today and Tomorrow.* Cambridge University Press, ISBN 978-0-521-20575-7.

Marshall C.A.M. *et al* (2023). Urban wild flower meadow planting for biodiversity, climate and society: An evaluation at King's College, Cambridge. *Ecological Solutions and Evidence* 4: 1–18.

Mata L., Amy K. Hahs A.K., Palma E., Backstrom A., Johnston N., King T., Olson A., R.,Renowden C., Smith R.,T., Vogel B., Ward S. (2023). Large positive ecological changes of small urban greening actions. *Ecological Solutions and Evidence* 4:3.

Mazzoleni S., Carteni F., Bonanomi G., Senatore M., Termolino P., Giannino F., Incerti G., Rietkerk M., Lanzotti V., Chiusano M.L. (2015). Inhibitory and toxic effects of extracellular self-DNA in litter: a mechanism for negative plant-soil feedbacks? *New Phytologist* 205:3, 1195–1210.

McCormick M.K. and Jacquemyn H. (2014). What constrains the distribution of Orchid populations? *New Phytologist* 202: 392–400.

McCormick M.K., Taylor D.L., Whigham D.F., Burnett R.K., (2016). Germination patterns in three terrestrial Orchids relate to abundance of mycorrhizal fungi. *Journal of Ecology* 104:3, 744–754.

Mendal J.G. (1866). Experiments in plant hybridization. *Proceedings of the Natural History Society of Brünn* IV: 3–47.

Merfield C. (2022). Redefining weeds for the post-herbicide era. *Weed Research* 62:4, 263–267.

Milligan C., Cox E., Alday J.G., Santana V.M., Mcallister H., Pakeman R.J., Le Duc., Marrs R. (2015). The effectiveness of old and new strategies for long-term control of *Pteridium aquilinum*, an 8-year test. *Weed Research* 53:3, 1–11.

Monkswood Review, (2005). A review of the potential effects of seed sowing for habitat re-creation on the conservation of intraspecific biodiversity. *Defra Report* CSG 15.

Montalvo A.M., and Ellstrand N.C. (2000). Transplantation of the subshrub *Lotus scoparius*: testing the home-site advantage hypothesis. *Conservation Biology* 14:4 1034-1045.

Montalvo A.M., and Ellstrand N.C. (2001). Nonlocal transplantation and outbreeding depression in the subshrub *Lotus scoparious* (Fabaceae). *American Journal of Botany* 88:2 258–269.

Mortimer E.J. (2016). Ecotypic variation in *'Lotus corniculatus L.'* and implications for grassland restoration: interaction of ecotypes with soil type and management, in relation to herbivory. *PhD thesis. Bath Spa University. ResearchsPAce.*SPECIAL FEATURE:

Mudrak O., Mladek J., Blazek P., Leps J., Dolezal J., Nekvapilova E., Tesitel J. (2014). Establishment of hemiparasitic Rhinanthus spp. in grassland restoration: lessons learned from sowing experiments. *Applied Vegetation Science* 17: 274–287

Nagel R., Durka W., Bossdorf O., Bucharova A., (2018). Rapid evolution in native plants cultivated for ecological restoration: not a general pattern. *Plant Biology* 21: 551–558.

Nichols R., Goulson D., Holland J., (2019). The best wild flowers for Wild Bees. *The Journal of Insect Conservation.* 23: 5–7.

Ohno T., Zibilske L.M., Liebman M., Gallandt E.R., Berube C. (2000). Phytotoxic effects of Red Clover amended soils on Wild Mustard seedling growth. *Elsevier. Agriculture Ecosystems and Environment* 78:1, 187–192.

Ohsaki H., Mukai H., Yamowa A. (2020). Biochemical recognition in seeds: Germination of *Rumex obtusifolius* is promoted by leaves of faciliative adult conspecifics. *Plant Species Biology* 35:3, 233–242.

Oldfield S. F., Olwell P., Shaw N., Havens K., (2019) Seeds of Restoration Success: Wild Lands & Plant Diversity in the U.S. *Springer Earth system Sciences* ISBN 978-3-319-96973-2.

Otto L.G, Junghams W.R., Plescher A., Sonnenschein, Sharbel T.F. (2015). Towards breeding of triploid chamomile (*Matricaria recutita* L.) – Ploidy variation within German chamomile of various origins. *Plant Breeding* 134:4, 485–493.

Pakeman R.J., Brooker R.W., Karley A.J., Newton A.C., Mitchell C., Hewison R.L., Pollenusa J., Guy D.C., Schob C. (2019). Increased crop diversity reduces the functional space available for weeds. *Weed Research* 60:2, 121–131.

Paril J. (2023). Heterosis in crop improvement. *The Plant Journal* 14:6, 1202–1506.

Parkins K.I., Clark J.A. (2015). Green roofs provide habitat for urban bats. *Global Ecology and Conservation* 4: 349–357.

Parra-Tabla V., Arceo-Gomez G., (2021). Impact of plant invasions in native plant-pollinator networks. *New Phytologist* 230:6, 2117–2128.

Pedrini S. and Dixon K.W. (2020). International principles and standards for native seeds in ecological restoration. *Restoration Ecology* 28:S3, 286–303.

Pedrini S., Gibson-Roy P., Trivedi C., Galvez-Ramirez C., Hardwick K., Shaw N., Frischie S., Laverack G., Dixon K. (2020). International principles and standards for native seeds in ecological restoratin. *Restoration Ecology* 28:S3, S286–303.

Pedrini S., Stevens J.C., Dixon K.W. (2021). Seed encrusting with salicylic acid: A novel approach to improve establishment of grass species in ecological restoration. *PLoS ONE* 16:6.

Peterken G. (2013). *Meadows.* Bloomsbury Publishing, ISBN: 9781472960344.

Polte S., Rheinhold K. (2013). The function of the Wild Carrot's dark central floret: attract, guide or deter? *Plant Species Biology* 28:1, 81–86.

Prillaman M. (2022). Are we in the Athropocene? Geologists could define new epoch for Earth *Nature* 613(7942), 14–15.

Pywell R., Bullock J. (2005). *A rattling good ruse.* Richard Pywell and James Bullock report on parasitic plants that help farmers restore wildflower Meadows. Centre for Ecology & Hydrology.

Pywell R., Bullock J.M., Tallowin J.R.B., Walker K.J., Warman E.A., Masters G. (2007). Enhancing the diversity of species-poor grassland: an experimental assessment of multiple constraints. *Journal of Applied Ecology* 44:1, 81–94.

Pywell R., Heard M.S., Bradbury R.S., Hinsley S., Nowakowski M., Walker K.J., Bullock J.M. (2012). Wildlife-friendly farming benefits rare birds, bees and plants. *Biology Letters* 8:5, 772–775.

Raabova J., Munzbergova Z., Fischer M. (2007). Ecological rather then geographic or genetic distance affects local adaption of the rare perennial herb, *Aster amellus. Biological Conservation* 139: 348–357.

Raguso R.A. (2023). Blood-stained flowers and bloodthirsty flies. *New Phytologist* 239:4, 1164–1165.

Redhead J. W., Sheail J., Bullock J.M., Ferreruela A., Walker K.J., Pywell R.F. (2013). The natural regeneration of calcareous grassland at a landscape scale: 150 years of plant community re-assembly on Salisbury Plain. *Applied vegetation Science* 17:3, 408–418.

Redhead J. W., Sheail J., Bullock J.M., Ferreruela A., Walker K.J., Pyewell R.F. (2022). The effects of a decade of agri-environment intervention in a lowland farm landscape on population trends of birds and butterflies. *Journal of Applied Ecology* 00: 1–11.

RHS (2015). Gardens as habitats for pollinators. *Plants for Bugs Bulletin.*

RHS Plant Finder (2022). Published by Royal Horticultural Society, ISBN: 9781911666233.

Rhys B., Jacquemyn H., Endels P., Van Rossum F., Hermy M., Triesst L., De Bryun L., Blust G.D.E. (2004). Reduced reproductive success in small populations of the self-incompatible *Primula vulgaris*. *Journal of Ecology* 92: 5–14.

Ridding L.E., Redhead J.W., Pywell R.F. (2013). Fate of semi-natural grassland in England between 1960 and 2013: A test of national conservation policy. Elsevier *Global Ecology and Conservation* 4: 516–525.

Riviere S., Provendier D., Malaval S., Sanson B., Gourvil J., Albert A., Millet J. (2022). Structuring supply chains of native plant material of wild and local provenance in France: A contribution to ecological restoration and nature-based solutions. Elsevier *Nature-Based Solutions* 2: 100035.

Rodwell J.S. (2006). *National Vegetation Classification: User's Handbook*. Published by Joint Nature Conservation Committee.

Rogers D.L., Montalvo A.M. (2004). Genetically appropriate choices of plant materials to maintain biological diversity. University of California, *USDA Forest Service.*

Rotches-Ribalta R., Blanco-Moreno J.M., Sans F.X. (2020). Reduced crop-sowing density improves performance of rare arable weed species more effectively than reduced fertilization. *Weed Research* 60:4, 269–277.

Rothero E., Lawson C. (2016). *Floodplain meadows – beauty and utility. A Technical Handbook.* Open University, ISBN 978-1-4370-2066-5.

Rottstock T., Kummer V., Fischer M., Joshi J. (2016). Rapid transgenerational effects in *Knautia arvensis* in response to plant-community diversity. *Journal of Ecology* 105:3, 714–725.

RSPB. (1997). The Wet Grassland Guide: Managing floodplain and coastal wet grassland for wildlife. *English Nature and Institute of Terrestrial Ecology.*

Ryan N., Laverack G., Powell A., (2008). Establishing quality control in UK wildflower seed production. *Seed Testing International* 135: 49–53.

Salisbury A., Armitage J.D., Bostock H., Perry J., Tatchell M., Thompson K. (2015). Enhancing gardens as habitats for flower-visiting aerial pollinators (insects): should we plant native or exotic species. *Journal of Applied Ecology* 52:5, 1156–1164.

Sanyal A., Decocq G. (2015). Biological Flora of the British Isles No. 277: *Crambe maritima*. *Journal of Ecology* 103:769–788.

Schmitz O.J., Sylvan M., Atwood T.B., Bakker E.B. (2023). Trophic rewilding can expand natural climate solutions. *Nature Climate Change* 13: 324–333.

Selosse M.A. (2014). The latest news from biological interactions in Orchids: in love head to toe. *New Phytologist* 202: 337–340.

Shantz H.L. and Piemiesel R.L. (1917). Fungus fairy rings in Eastern Colorado and their effect on vegetation. *Journal of Agricultural Research* No.5 191–245.

Sharifi R. and Ryu C. (2020). Social networking in crop plants: Wired and wireless cross-plant communications. *Plant Cell & Environment* 44:4, 1095–1110.

Shaw N., Barak R.S., Campbell R.E., Kirmer A., Pedrini S., Dixon K., Frichie S. (2020). Seed use in the field: delivering seed for restoration success. *Restoration Ecology* 28:S3, 276–285.

Silvertown J., Poulton P.R., Johnston A.E., Edwards G., Heard M., Bliss P.M. (2006). The Park Grass Experiment 1856-2006: its contribution to ecology. *Journal of Ecology* 94: 801–814.

Skilbeck C.A. (2023). Achene dispersal in Asteraceae with specific reference to British and Irish *Centaurea*. *British & Irish Botany* 5:1, 69–87.

Smith B., Diaz A., Daniels R., Winder L., Holland J.M. (2007). Regional & ecotypes traits in *Lotus corniculatus* L., with Reference to restoration ecology. *Restoration Ecology* 17:1, 12–23.

Smith J.D., Johnson B.I., Mescher M.C., De Moraes C.M. (2020). A plant parasite uses light cues to detect differences in host-plant proximity and architecture. *Plant, Cell and Environment* 44:4, 1142–1150.

Smith M.R., Mueller N.D., Springmann M., Sulser T.B., Garibaldi L.A., Gerber J., Wiebe K., Myers S.S. (2022). Pollinator deficits, food consumption, and consequences for human health: a modelling study. *Environmental Health Perspectives* 130:12, 1–12.

Smith R.A.H. and Bradshaw A.D. (1979). The use of metal tolerant plant populations for the reclamation of metalliferous wastes. *Journal of Applied Ecology* 16: 595–612.

Smith R.S., Shiel R.S., Millward D., Corkhill P., Sanderson R.A., (2002). Soil seed banks and the effects of meadow management on vegetation change in a 10-year meadow field trial. *Journal of Applied Ecology* 39: 279–293.

Smith S.M. (2020). Seasonal control of seed germination. *New Phytologist* 225:5, 1821–1823.

Spencer J. and Kirby K. (1992). An inventory of woodland for England and Wales. *Biological Conservation.* 62: 77–93.

Stace C.A. (2019). *New Flora of the British Isles.* C&M Floristics, ISBN: 9781527226302.

Stace C.A. (2022). *Concise Flora of the British Isles.* C&M Floristics, ISBN: 9781399919609.

Star-Keddle R.E. (2022). Evaluating the success of upland hay meadow restoration in the North Pennines, United Kingdom, using green hay transfer. *Ecological Solutions and Evidence* 3:1, 1–11.

St Claire A.B., Dunwiddie P., Fant J.B., Kaye T.N. (2022). Mixing source populations increases genetic diversity of restored rare plant populations. *Restoration Ecology* 28:3, 1–11.

Steele D.J., Baldock K.C.R., Breeze T.D., Brown M.J.F.,Carvell C., Dicks L.V., Garratt M.P., Norman H., Potts S.G., Senapathi D., Vanbergen A.J. (2019). *Management and drivers of change of pollinating insects and pollination services.* Central Archive @ University of Reading. Published by DEFRA.

Stephen K. (1996). *Worcestershire grasslands – report of botanical survey.* Worcestershire Wildlife Trust.

Stevens C. J., Wilson J., McAllister H.A. (2012). Biological Flora of the British Isles: *Campanula rotundifolia*. *Journal of Ecology* 100:1, 821–839.

Stokstad E. (2023). Unmixed blessing. Researchers are closing in on the long-sought goal of creating high-yielding hybrid crops that can be propagated indefinitely, without sex. *Science* 19:380, 684–687.

Storkey J., Neve P., Liebman M. (2018). What good is weed diversity? *Weed Research* 58:4, 239–243.

Stroh P.A., Walker K.J., Humphrey T.A., Pescott O.L., Burkmar R.J., (2023). *Plant Atlas 2020: mapping changes in the distribution of the British and Irish Flora.* Princetown University Press. ISBN: 9780691247595.

Stroh P.A. (2015a). Green-winged Orchid, *Anacamptis morio* L., *Species Account BSBI.*

Stroh P.A. (2015b).Dyers greenwood, *Genista tinctoria* L., *Species Account BSBI.*

Stroh P.A. (2015c). Meadow Saxifrage, *Saxifraga granulata* L., *Species Account BSBI.*

Stroot L., Brinkert A., Holzel N., Rusing A. (2021). Establishment of Wildflower strips in a wide range of environments: a lesson from a landscape-scale project. *Restoration Ecology* 30:4, 13542.

Tatarenko I., Walker K., Dyson M. (2022). Biological Flora of Britain and Ireland: *Fritillaria meleagris. Journal of Ecology*: 110, 6 1–23.

Taylor K. (2009). Biological Flora of the British Isles: *Urtica dioica* L., *Journal of Applied Ecology* 97:6, 1436–1458.

Thomas G.J., Allen D., Grose M. (1981). The demography and flora of the Ouse Washes, England. *Biological Conservation* 21: 197–229.

Toit J.T. and Petorelli N. (2019). The differences between rewilding and restoring an ecologically degraded landscape. *Journal of Applied Ecology* 56:11, 2467–2471.

Tuberlin A.J., Malamud B.D., Frances R.A. (2017). Mapping the global state of invasive alien species: patterns of invasion and policy responses. *Global Ecology and Biogeography* 26: 78–92.

Turreson G. (1922). The Species and the Variety as Ecological Units. Institute of Genetics, Sweden. *Hereditas* 100–113.

UK Government (2018). *A Green Future: Our 25 Year Plan to Improve the Environment.* Published by DEFRA.

UK Government (2021). *Environment Act 2021.* Published by DEFRA.

UK Government (2023). *Environmental Improvement Plan 2023.* Published by DEFRA.

Valerio-Galan M., Ibanez-Gaston R., Gotzenberger L., (2022). Long-term year-to-year stability and its drivers in a Mediterranean grassland. *Journal of Ecology* 110:5, 1174–1188.

Varga S., Soulsbury C.D., John E.A. (2022). Biological Flora of Britain and Ireland: *Knautia arvensis. Journal of Ecology* 110:8, 1970–1992.

Veits M., Khait I., Obloski U., Zinger E., Boonman A., Goldshtein A., Saban K., Ben-Dor U., Estlein P., Kabat A., Peretz D., Ratzersdorfer I., Krylov S., Chamovitz D., Sapir Y., Yuvel S. Hadnay L. (2019). Flowers respond to pollinator sound within minutes by increasing nectar sugar concentration. *Ecology Letters* 22:9, 1483–1492.

Villermereuil P., Gaggiotii O.E., Mouterde M., Till-Bottraud I., (2015). Common garden experiments in the genomic era: new perspectives and opportunities. *Heredity* 116: 249–254.

Villiers T.A. (1974). Genetic maintenance of seed in imbibed storage. In *Crop genetic resources for today and tomorrow.* Edited by Frankel O.H. & Hawkes J.G. ISBN: 9780521179706.

Vitis de M., Hay F.R., Dickie J.B., Trivedi C., Choi J., Fiegner R. (2020). Seed storage. Maintaining seed viability and vigor for restoration use. *Restoration Ecology* 28:S3, 249–255.

Volis S., Blecher M. (2010). Quasi in situ: a bridge between ex situ and in situ conservation of plants. *Biodiversity Conservation*: 19, 2441–2454.

Voller E., Bossdorf O., Prati D., Harald A. (2017). Evolutionary responses to land use in eight common grassland plants. *Journal of Ecology* 105:5, 1290–1297.

Wagner M., Hulmes S., Hulmes L., Redhead J.W., Nowakowski M., Pywell R.F. (2020). Green hay transfer for grassland restoration: species capture and establishment. *Restoration Ecology* 29:S1 13259

Walker K. (2021). Snake's-head Fritillary *Fritillaria meleagris* (Liliaceae) in Britain: its distribution, habitats and status. *British & Irish Botany* 3:3, 263–278.

Walker K.J., Hodder K.H., Bullock J.B., Pywell R.F. (2004). *A review of the potential effects of seed sowing for habitat re-creation on the conservation of intraspecific biodiversity.* Defra Contract BD1447. Centre for Ecology and Hydrology.

Waring P. (1990). Observations on invertebrates collected up during wild flower seed harvesting in a hay meadow, with particular reference to the butterflies and moths. *British Journal of Entomology and Natural History* 3:1, 143–152.

Watson J.D. and Crick F.H.C. (1953). A structure for deoxyribose nucleic acid. *Nature* 4356: 737–738.

Waud M., Wiegand T., Brys R., Lievens B., Jacquemyn H. (2016). Non-random seedling establishment corresponds with distance-dependent decline in mycorrhizal abundance in two terrestrial orchids. *New Phytologist* 211:1, 255–264.

Wells T.C.E. & Cox, R. (1991). Demographic and biological studies on *Ophrys apifera*: some results from a 10-year study, in Wells, T. C. E. & Willems, J. H., eds. *Population ecology of terrestrial orchids,* 47–63.

Wells T.C.E., Bell S., Frost A. (1981). *Creating attractive grasslands using native plant species.* Nature Conservancy Council.

Westbury D.B., Harris S.J., Brown V.K., Potts S.G. (2018). The influence of seed mix and management on the performance and persistence of sown forbs in buffer strips. *Aspects of Applied Biology* 139: 1–7.

Westbury D.B., Davies A., Woodcock B.A., Dunnett N.P. (2006). Seeds of change: The value of using *Rhinanthus minor* in grassland restoration. *Journal of Vegetation Science* 17:4, 435–446.

Westbury D.B. (2004). *Rhinanthus minor* L. Biological Flora of the British Isles. *Journal of Ecology* 92:5, 906–927.

Whigham D.F., O'Neill J.P., Rasmussen H.N., Caldwell B.A., McCormick M.K. (2006). Seed longevity in terrestrial Orchids – potential for persistent *in situ* seed banks. *Biological Conservation* 129: 24–40.

Woodcock B.A., McDonald A.W., Pyewell R.F. (2011). Can long-term floodplain meadow recreation replicate species composition and functional characteristics of target grasslands? *Journal of Applied Ecology* 48:5, 1070–1078.

Worede M. et al. (2009). Interview. *Seeding* 16–19.

Zotti M., Cesarano G., Ercolini D., Tessi G., Allegrezza M., Giannino F., Mazzoleni S., Bonanomi G. (2020). One ring to rule them all: an ecosystem engineer fungus fosters plant and microbial diversity in a Mediterranean grassland. *New Phytologist* 227:3, 884–898.

INDEX

Achillea millefolium 95, **150**, 254
Achillea ptarmica **150**, 254
Adonis annua **150**, 254
Aethusa cynapium **150**, 254
agri-environment schemes 54
Agrimonia eupatoria 49, 97, **151**, 254
Agrostemma githago 90, **152**, 254
Agrostis capillaris 29, 106, **152**, 254
Agrostis stolonifera 29
Ajuga reptans **152**, 254
alien species 66
Alliaria petiolata **153**, 254
Allium ursinum 97, **153**, 254
Alopecurus geniculatus **154**, 254
Alopecurus myosuroides 87, 90
Alopecurus pratensis **154**, 254
Anacamptis morio 95, **154**, 254
Anacamptis pyramidalis 95, **155**, 254
Anemone nemorosa 109, **155**, 254
Angelica sylvestris **156**, 254
Anisantha tectorum 31–32
Anthoxanthum odoratum 68, 107, 119, **156**, 254
Anthriscus sylvestris **157**, 254
Anthyllis vulneraria 137, **157**, 254
apomictic species 11, 70
archaeophyte species 12
Armeria maritima **158**, 254
Arrhenatherum elatius 29, 119, **158**, 254
Arum maculatum **159**, 254
Asplenium scolopendrium 138, **159**, 254
Avena fatua 90
Avenella flexuosa **160**, 254
Avenula pubescens **160**, 254

Barbarea vulgaris **160**, 254
Bellis perennis **160**, 254
Betonica officinalis 78, **161**, 254
Blackstonia perfoliata **162**, 255
biodiversity 19
Brachypodium sylvaticum **162**, 255
Briza maxima 29, 33
Briza media **162**, 255
Bromopsis erecta 133, **163**, 255
brush harvesting 16, 100–103

Calluna vulgaris **163**, 255
Caltha palustris **164**, 255
Campanula glomerata **164**, 255
Campanula rotundifolia **165**, 255
Campanula rapunculoides 57
Campanula trachelium **165**, 255
Capsella bursa-pastoris 87, **166**, 255
Cardamine pratensis **166**, 255
Carduus nutans **167**, 255
Carex divulsa **167**, 255
Carex echinata **168**, 255
Carex flacca **168**, 255
Carex pendula 138, **169**, 255
Carex sylvatica **169**, 255
Carlina vulgaris **169**, 255
Centaurea cyanus 90, **170**, 255
Centaurea nigra 33, 42, 49, 68, 76, 95, 111, 133, **171**, 255
Centaurea scabiosa **171–172**, 255
Centaurium erythraea **172**, 255
Cerastium fontanum **172**, 255
Chaenorhinum minus **173**, 255
Chaerophyllum temullum **173**, 255
chemical farming 52
Cichorium intybus 29
Circaea lutetiana **174**, 255
Cirsium acaule **174**, 255
Cirsium arvense 35, 76, **125**
Cirsium eriophorum **174**, 255
Cirsium palustre **175**, 255
Cirsium vulgare 35, 76, 121, **125**
Clematis vitalba **175**, 255
climate moderation 41
Clinopodium vulgare **176**, 255
Conium maculatum 127
Conopodium majus **176**, 255
Cotoneaster horizontalis 32
Crambe maritima **177**, 255
creation method 14, 82
Crithmum maritimum **177**, 255
Cruciata laevipes **177–178**, 256
cultivars 66
Cynoglossum officinale **178**, 256
Cynosurus cristatus 95, **178**, 256

Dactylis glomerata 29, 119, **179**, 256
Dactylorhiza fuchsii 12, 21, 71, 95, 98, 146, **179**, 256
Dactylorhiza praetermissa 71, 95, **180**, 256
Daucus carota 28, 29, 68, 95, **180**, 256
Deschampsia cespitosa **181**, 256
Digitalis purpurea 97, **182**, 256
Dipsacus fullonum **182**, 256
Dryopteris felix-mas **183**, 256

Echium vulgare **183–184**, 256
ecosystem services 41–45
ecotypes **22–23, 67–68**
Epipactis heleborine 95, **184**, 256
Epipactis palustris 95, **184**, 256
Erica cinerea **184**, 256
Erica tetralix **185**, 256
Erigeron karvinskianus 58
Ervilia sylvatica **185**, 256
Eryngium maritimum **185**, 256
Eupatorium cannabinum **185**, 256
Euphorbia exigua **186**, 256
Euphorbia helioscopia **186**, 256
Euphrasia sp 83–84

Festuca ovina 29, **186**, 256
Festuca rubra 29, 71, 95, 106, **187**, 256
Filipendula ulmaria 48, 119, **187–188**, 256
Filipendula vulgaris **188**, 256
Fragaria vesca **188**, 256
Fritillaria meleagris 12, 13, 33, 48, 66, **189**, 256
Fumaria officinalis 101, **189**, 256
fungi 144

Galeopsis tetrahit **190**, 256
gene-flow 49–50, 70
genetic drift 70
genetic rescue 50
genetic swamping 71
Galium album 20, **190**, 256
Galium odoratum **190**, 256
Galium verum 63, 95, 119, **191**, 256
Genista tinctoria **192**, 256
Geranium dissectum **192**, 256
Geranium lucidum **193**, 257
Geranium molle **193**, 257
Geranium pratense 12, 49, 111, **193–195**, 257
Geranium pyreniacum 12, 66, **195**, 257
Geranium robertianun **195**, 257
Geum rivale **196**, 257
Geum urbanum **196**, 257
Glaucium flavum **196**, 257
Glebionis segetum 129, **197**, 257
grazing 121–124

Helianthemum nummularium **197**, 257
Heracleum sphondylium 76, **198**, 257
heterosis 71
Himantoglossum hircinum 26–27, 60, 70
Hippocrepis comosa **198**, 257
Hordeum secalinum **199**, 257
Hordeum vulgare 52–53
Hyacinthoides non-scripta 25, 27, 58–59, 142, **199**, 257
Hypericum perforatum 33
Hypochaeris radicata 97, **200**, 257

inbreeding depression 71
introgression 71
invasive aliens 31
Iris foetidissima **200**, 257
Iris pseudacorus 97, **201**, 257

Jacobea vulgaris 35, 121–122, **125–127**
Jasione montana **201**, 257

Kickxia elantine **202**, 257
Kickxia spuria **202**, 257
Knautia arvensis 44, 106, **202**, 257
Koeleria macrantha **203**, 257

Lamiastrum galeobdolon 142, **203**, 257
Lamium album **203**, 257
Lamium purpureum **204**, 257
Lapsana communis **204**, 257
Lathyrus latifolis 58
Lathyrus pratensis **204**, 257
Lathyrus sylvestris **205**, 257
Legousia hybrida **205**, 257
Leontodon hispidus 49, 122, **206**, 257
Leucanthemum vulgare 66, 76, 95, 111, 119, 130, 132, **206–207**, 257
Linaria vulgaris **207**, 257
Linum bienne **207**, 257
Linum catharticum **208**, 258
local ecotype 67–69
Lolium multiflorum 87
Lolium perenne 29, 52, 71, 87
Lotus corniculatus 17, 20, 29, 49, 67, 76, 95, **208**, 258
Lotus pedunculatus **209**, 258
Lycopus europeus **209**, 258
Lysimachia arvensis 87, **209–210**, 258
Lythrum salicaria 140, **210**, 258

Malva moschata **211**, 258
Malva sylvestris 68, **211**, 258
Matricaria chamomilla **212**, 258
Matricaria discoidea **212**, 258
Medicago lupulina 29, **212**, 258
Mentha aquatica **213**, 258
Mercurialis perennis **213–214**, 258
micro species 22
mowing dates 119–121
Myosotis arvensis **214**, 258

native species 11, 66
natural capital 39–40
natural regeneration 76, 79
naturalised neophyte species 12
Neottia ovata 95, **214**, 258
non-naturalised neophytes 13
nurse species 86

Odontites vernus **215**, 258
Oenanthe pimpinelloides **215**, 258
Onobrychis viciifolia 29, **216**, 258
Ononis repens **216**, 258
Ophioglossum vulgatum **217**, 258
Ophrys apifera 20, 66, 70, 95, 145, **217–219**, 258
orchids 49, 61, **95**, 98, 144–147
Orchis mascula 95, **219**, 258
Origanum vulgare 42, 63, 119, **219**
origin 65
Orobanche minor 20, 145, **220**, 258
outbreeding depression 71

Papaver dubium **220**, 258
Papaver rhoeas 12, 53, 66, 110, 129–130, **221**, 258
Paris quadrifolia **221**, 258
Pastinaca sativa **222**, 258
Phleum arenarium **222**, 258
Phleum bertolonii 29, **222**, 258
Phleum pratense 29
Pilosella officinarum **223**, 258
Pimpinella major **223**, 258
Pimpinella saxifraga **223**, 258
Plantago coronopus **224**, 258
Plantago lanceolata 95, **224–225**, 259
Plantago media **225**, 259
Plantanthera chlorantha 95, **225**, 259
plant ID 14
pollinator services 41–44
Polygala vulgaris **226**, 259
Polypodium vulgare **226**, 259
Poterium sanguisorba 76, **226–227**, 259
Primula veris 49, 66, 70, 95, 104–106, 111, 122, 133–135, **227–228**, 259
Primula vulgaris 58, **228**, 259
proscribed species 28
protected species 25
provenance 65
Prunella vulgaris 95, **229**, 259
Pulicaria dysenterica **229–230**, 259
Pteridium aquilinum 36–37, 81

Ranunculs acris 43, 95, **230**, 259
Ranunculus arvensis **231**, 259
Ranunculus auricomus **231**, 259
Ranunculus bulbosus 119, 130, **231–232**, 259
Reseda lutea **232**, 259
restoration method 14, 81–82
restoration species 91, **149–253**, **254–260**
Rhododendron ponticum 12, 32
Rhinanthus minor 12, 81, **83–85**, 95, 111, 114, 119, 132–133, **232–233**, 259
Rubia perregrina **233**, 259
Rumex acetosa **233**, 259
Rumex acetosella **234**, 259
Rumex crispus 35, 76, **124**
Rumex obtusifolius 35, 76, **124**

Salvia pratensis 25, 27
Salvia verbenaca **234**, 259
Sanguisorba officinalis 48, 119, **235**
Sanicula europeus **235**, 259
Saxifraga granulata **236**, 259
Scabiosa columbaria 76, **236**, 259
Scrophularia auriculata **236**, 259
Sedum acre 63, **237**, 259
seed cleaning 109
seed dormancy 111
seed germination 111, 116
seed harvest 106–107
seed origin 65
seed provenance 65
seed rates **114**, **254–260**
seed storage 110
seed viability 111
Senecio squalidus 12
Senecio vulgaris 87
sensitive species 27
Serratula tinctoria 119, **237**, 259
Sherardia arvensis **237–238**, 259
Silaum silaus 48, **238**, 259
Silene dioica 70, **238**, 259
Silene flos-cuculi **239**, 259
Silene latifolia 66, **239**, 259
Silene uniflora **240**, 259
Silene vulgaris **240**, 259
Sinapis arvensis 87
species 20
Spiranthes spiralis 95, **241**, 260
stock collection 103–105
Stachys sylvatica **241–242**, 260
Stellaria holostea **242**, 260
Succisa pratensis 119, **242**, 260

Taraxacum officinale 22, 34, 49, **54**, 70, 97, 119, 130, **243**, 260
Teucrium scorodonia **243–244**, 260
Thalictrum flavum **244**, 260
Thymus drucei **244**, 260
Torilis japonica **245**, 260
Tragopogon pratensis **245**, 260
Trifolium campestre **245**, 260
Trifolium dubium **246**, 260
Trifolium pratense 29, **246**, 260
Trifolium repens 29, **247**, 260
Trisetum flavescens **247–248**, 260
Triticum aestivum 52–54
Typha latifolia **248**, 260

Umbilicus rupestris **248**, 260
Urtica dioica 43, 81, 121, **127**

Verbascum nigrum **249**, 260
Verbascum thapsus **249**, 260
Veronica chamaedris **250**, 260
Veronica persica 12, 89, **250**, 260
Vicia cracca **250**, 260
Vicia sativa 29, **251–252**, 260
Vicia sepium **252**, 260
Viola arvensis **252**, 260
Viola riviniana **253**, 260
Viola tricolor **253**, 260

weeds 31–37, 76–77, 124–127
wilding 79
wild type 66

First published in 2024 by
The Crowood Press Ltd
Ramsbury, Marlborough
Wiltshire SN8 2HR

enquiries@crowood.com

www.crowood.com

British Library Cataloguing-in-Publication Data
A catalogue record for this book is available from the British Library.

ISBN 978 0 7198 4438 6

The right of Donald MacIntyre to be identified as author of this work has been asserted by him in accordance with the Copyright, Designs and Patents Act 1988.

Cover design by Sergey Tsvetkov

Typeset by SJmagic DESIGN SERVICES, India
Printed and bound in India by Nutech Print Services

Acknowledgements

I have so many people to thank. I would not have followed this path had my dear mother Mary not inspired us to go Orchid hunting in the Somerset meadows. And so began a fascination with, and an appreciation of, the beauty of the wild and wild plants that grew over many years and featured in many plant-hunting forays. On leaving school I followed the botanical route and was inspired along the way by Mrs Birch (Botany tutor at Bath Technical College), Dr Peter Dixon (undergraduate Botany tutor at Royal Holloway College, University of London), Dr Charles Elliott (post-graduate Genetics tutor at Glasgow University), Dr Terry Wells (researcher at Monks Wood Experimental Station), and many others. With thanks to all.

I have been ably assisted by Jane Lipington in the preparation of this book. Jane compiled first drafts of the species accounts, the reference list and the glossary, and commented on drafts of the main text. Draft texts were also seen by Dr Sarah Oldfield, Dr Barbara Smith, Ann Skinner, Sue Garratt, Dr Colin Pope, Grace Macmillan and Hannah Gardner, all of whom contributed many good ideas.

Myrica MacIntyre prepared figures 1 and 2 in Chapter 2. Images were provided by Charlie Carter, p.63 (middle); Chris Park, p.139 (bottom); David Trudgill, p.61; Dr Erica Mortimer, p.67; Emorsgate Seeds, p.136 (top right); Joshua Day, p.115 (top); Laurie MacIntyre, p.100 (top right), p.236 (top right), p.242 (top right); Marc Bullock, p.107 (top left), p.110; and Marianne Pickford, p.99 (bottom right), p.121 (right). The remaining images are those of the author. Many thanks to all those who have contributed.